# FINAL OPERATING REPORT

## FOR

# PM-3A NUCLEAR POWER PLANT McMURDO STATION, ANTARCTICA

Prepared by:

## U. S. NAVAL NUCLEAR POWER UNIT

### P. O. BOX 96

### FORT BELVOIR, VIRGINIA-22060

Report Number 69

12 MARCH 1964—20 OCTOBER 1973

NAVAL NUCLEAR POWER UNIT
P. O. BOX 96
FORT BELVOIR, VIRGINIA 22060

# PM-3A NUCLEAR POWER PLANT
## FINAL OPERATING REPORT

Prepared by:
Plans and Operations Department
Naval Nuclear Power Unit

Approved by:

G. E. KRAUTER
Commander
Civil Engineer Corp
United States Navy
Officer in Charge

## PREFACE

This report provides historical information
relating to the operation, maintenance and technical support of the PM-3A Nuclear Power Plant and
the Seawater Distillation Plant, located at McMurdo
Station, Antarctica. This final report is an accumu-
lation of the information gathered over the ten and
one-half years of Naval operation and also provides
the background information which has lead to the
decision to decommission and remove the PM-3A
from Antarctica.

The report is prepared by the Naval Nuclear
Power Unit, P. O. Box 96, Fort Belvoir, Virginia
22060. Comments are invited and should be for-
warded to the Naval Nuclear Power Unit.

# TABLE OF CONTENTS

# LIST OF TABLES

# LIST OF FIGURES

# SECTION I

## LIFETIME SUMMARY OF PM-3A OPERATIONS

Project Initiation. During the 1960 Congressional Hearings, it was determined that the construction of nuclear power plants in Antarctica would cut the cost of operations, particularly logistics, at our stations there. As a result of these hearings, Congress authorized and made available funds to construct the PM-3A Nuclear Power Plant in Antarctica at McMurdo Station. The U. S. Atomic Energy Commission, in August 1960, initiated work by fixed-price contract with the Martin Company for the design, fabrication, preshipment testing, packaging, transportation, installation and on-site testing and initial operation of the first nuclear power plant for Antarctica. The design capacity of the plant was 1800 KWE at 0.8 power factor.

Plant Installation. A tight schedule was maintained to allow factory pre-assembly and testing of the PM-3A prior to its being shipped to Antarctica during late 1961. The plant arrived at McMurdo Station aboard the USNS ARNEB on 12 December 1961 and was erected by Mobile Construction Battalion ONE and the Navy startup crew under the supervision of the Martin Company.

Initial Criticality and Navy Operation. The first nuclear criticality was achieved on 3 March 1962 and the first usable electrical power was supplied to McMurdo Station on 10 July 1962. Since that time the PM-3A has supplied 60,938,800 KWHRs of electrical energy to McMurdo Station. The Navy was authorized to operate the PM-3A on 27 May 1964. After completion of the necessary testing the first electrical energy supplied under Navy operation was produced on 10 June 1964. Since that date the PM-3A has been available to supply power to McMurdo Station for 54,182 hours 49 minutes out of a possible 74,976 hours for an availability of 72.266%. Details of the PM-3A operating data are contained in Section II of this report. The previous data values given were for the period of Navy operations ending 30 December 1972. The PM-3A was shutdown during calendar year 1973 pending the resolution of the interconnect leak and chloride stress corrosion problems. Details of these problems may be found in Section VIII of this report. In October the decision was finalized to remove the plant.

SECTION II

PLANT DATA

A. CHRONOLOGICAL HISTORY OF PLANT OPERATIONS

The following is a chronological history of PM-3A operations, maintenance and modifications for the period of 12 March 1964 through 20 October 1973 inclusive.

| Date | Event |
|---|---|
| 12 March 1964 | Plant placed in custody of the U. S. Navy in a shutdown maintenance status pending the resolution of safety problems. |
| 05 May 1964 | Received final AEC concurrence on site corrective action to control rod drive mechanism buffer pistons and actuator cylinders and that six units are completely satisfactory for operation. |
| 27 May 1964 | Navy operation authorized for test and evaluation. |
| 01 June 1964 | Precritical tests completed and reactor brought critical for core physics tests. |
| 06 June 1964 | Transient noise problems encountered in scram circuitry causing spurious plant scrams. PM-3A Operating Report Number 1; Malfunctions 64-27, 64-28. |
| 07 June 1964 | Completed core physics tests and proceeded to power operations. PM-3A Operating Report Number 1; Malfunctions 64-29, 64-30. |
| 10 June – 18 June 1964 | Plant initially picked up McMurdo Station heater load; however, plant operations were eventually limited to carrying plant load. Transients in reactor outlet temperature scram circuitry prevented assumption of McMurdo Station load. PM-3A Operating Report Number 1; Malfunctions 64-32, 64-33, 64-34, 64-35, 64-36, 64-37, 64-38, 64-39. |

2

| | |
|---|---|
| 18 June 1964 | Transient problems temporarily relieved by installation of capacitors and switches on outputs of high reactor outlet temperature scram bistables. |
| 18 June – 27 June 1964 | Plant carried McMurdo Station load for 205 hours of 240 hours total. PM-3A Operating Report Number 1; Malfunctions 64-39, 64-42, 64-43. |
| 28 June – 01 July 1964 | Plant carried McMurdo Station load. |
| 01 July 1964 | Reactor scrammed during bistable drift tests. Plant returned to power 9 hours after scram. PM-3A Operating Report number 2; Malfunction 64-44. |
| 01 July – 11 July 1964 | Plant carried McMurdo Station load. |
| 11 July 1964 | Reactor scrammed during scram logic tests. Plant returned to power 9 hours after scram. PM-3A Operating Report Number 2; Malfunction 64-45. |
| 11 July – 15 July 1964 | Plant carried McMurdo Station load. |
| 15 July 1964 | Reactor scrammed during scram logic tests and was returned to critical operations in one hour. A second reactor scram occurred due to pump power while placing condenser fan in operation. Plant returned to power within 8 hours of initial scram. PM-3A Operating Report Number 2; Malfunctions 64-46, 64-47. |
| 15 July – 20 July 1964 | Plant carried McMurdo Station load. |
| 20 July – 21 July 1964 | Plant shutdown on planned schedule for maintenance and modification. |
| 22 July – 29 July 1964 | Plant carried McMurdo Station load. |

| | |
|---|---|
| 29 July – 04 August 1964 | Plant shutdown by manual scram due to pressurizer heater failure. PM–3A Operating Report Number 2; Malfunction 64–48. |
| 04 August 1964 | Plant returned to power. |
| 04 August – 08 August 1964 | Plant carried McMurdo Station load. |
| 08 August 1964 | Reactor scrammed due to faulty containment pressure switch when the switch was inadvertently disturbed. Plant returned to power 12 hours after scram. PM–3A Operating Report Number 3; Malfunction 64–49. |
| 09 August – 17 August 1964 | Plant carried McMurdo Station load. |
| 17 August 1964 | Reactor scrammed while preparing for planned shutdown to change resins in the primary demineralizer. PM–3A Operating Report Number 3; Malfunction 64–51. |
| 17 August – 20 August 1964 | Plant shutdown for primary demineralizer resin change since chemistry data indicated continued upward trend in activity level of the primary coolant. PM–3A Operating Report Number 3; Malfunction 64–50. |
| 20 August – 22 August 1964 | Plant returned to power and carried McMurdo Station load. |
| 22 August 1964 | Reactor scrammed when containment pressure switch was disturbed. A second scram occurred during startup in the intermediate range when the condenser fan was energized in reverse direction, and a low electrical power frequency resulted. Plant returned to power in 13 hours, 20 minutes of initial scram. PM–3A Operating Report Number 3; Malfunctions 64–52, 64–53. |
| 23 August 1964 | Reactor scrammed due to transients in instrumentation; probable cause was broken lead in PC pump power converter. Plant was at power for 6.5 hours preceeding the scram and returned to power 8.5 hours after scram occurred. PM–3A Operating Report Number 3; Malfunction 64–54. |

4

| | |
|---|---|
| 23 August – 04 September 1964 | Plant carried McMurdo Station load. |
| 04 September 1964 | Planned shutdown due to high primary system iodine. Reactor scrammed while transferring plant electrical load to Caterpillar diesel generator on planned shutdown. Scram due to voltage and frequency swings. PM-3A Operating Report Number 4; Malfunctions 64-55, 64-56. |
| 04 September – 07 September 1964 | Plant down for replacement of source range detectors. PM-3A Operating Report Number 4; Malfunctions 64-57, 64-58. |
| 08 September – 11 September 1964 | Plant carried McMurdo Station load. |
| 11 September 1964 | Reactor manually scrammed due to the loss of control rod number 3 while at power. PM-3A Operating Report Number 4; Malfunction 64-59. |
| 11 September – 13 September 1964 | Plant down for replacement of both source range detectors. PM-3A Operating Report Number 4; Malfunction 64-60. |
| 14 September 1964 | Plant returned to power. |
| 14 September – 09 October 1964 | Plant carried McMurdo Station load. |
| 09 October 1964 | Plant dropped load for less than 2 hours for Crew IV training. |
| 09 October – 26 October 1964 | Plant carried McMurdo Station load. |
| 26 October – 31 October 1964 | Plant shutdown for source range detector replacement, nuclear instrumentation testing, primary and shield water demineralizer resin changes and Crew IV training. PM-3A Operating Report Number 5; Malfunctions 64-69, 64-70, 64-71. |

| 31 October 1964 | Control rod number 3 failed to drive up during operation to bring the reactor critical and the reactor was manually scrammed. PM-3A Operating Report Number 5; Malfunction 64-72. |

31 October 1964 — Reactor brought critical and plant load carried.

01 November 1964 — Reactor scrammed during planned shutdown for training due to switching transient. A second reactor scram occurred due to test transient during bistable trip test. PM-3A Operating Report Number 6; Malfunctions 64-73, 64-75.

01 November – 04 November 1964 — Plant shutdown for crew training and engineering support team work effort. Reactor brought critical and scrammed for training. Plant load picked up and dropped for training. Engineering support team tests in progress.

04 November – 07 November 1964 — Plant up for crew training and engineering support team work effort. Plant carried McMurdo Station load, maintaining full power with bypass steam for 10 hour xenon transient test. Steam generator chemistry and radiochemistry out of specifications due to system cycling during training on 5 November. Source range channel number 2 malfunctioned on 6 November due to a loose connector on the drawer. High air activity was encountered in the primary building on 7 November due to excessive number of iodine samples. PM-3A Operating Report Number 6; Malfunctions 64-76, 64-77, 64-78.

08 November 1964 — Reactor scrammed manually from 100% load for operator training. Hot rod drop tests and temperature coefficient runs performed at this time.

08 November – 11 November 1964 — Plant down for testing and maintenance.

| 11 November –<br>13 November 1964 | Reactor brought critical for engineering support team work effort. Continued scram response tests and performed cold rod drop tests. |
| :-- | :-- |
| 13 November 1964 | Reactor shutdown and placed in cold iron condition for scheduled summer maintenance and modification. |
| 14 November –<br>19 November 1964 | Purging containment for maintenance and modification. High air activity in containment forced delay in purging. PM-3A Operating Report Number 6; Malfunction 64-80. |
| 20 November 1964 | Containment purge completed, containment opened. Plant modifications and maintenance program initiated. |
| 21 November 1964 | Removal of shield water to temporary storage tanks under air blast condenser started. |
| 23 November 1964 | CRDM and position indicator cans removed. |
| 26 November 1964 | Transfer of shield water complete; sluice gate, dolly and old fuel transfer equipment removed. |
| 30 November 1964 | Installation of new fuel transfer equipment completed. |
| 01 December –<br>05 December 1964 | Hairline cracks rewelded in stainless steel liners of reactor and spent fuel tanks and refueling interconnect. Instrumentation logic test drawer switches replaced and new core installation monitoring system checked and tested. PM-3A Operating Report Number 7; Annex I. |
| 06 December –<br>07 December 1964 | Returned shield water to containment tank from temporary storage area. Installed temporary shield water filtering system. Rebuilt spent fuel cask to proper dimensions for spent core tank storage. |
| 08 December –<br>09 December 1964 | Placed spent fuel cask in spent fuel tank. Cleaned up primary building and made final tool check in preparation for refueling. |
| 10 December 1964 | Completed defueling spent core. Placed spent core in spent fuel cask at 102240 (local time). |

| 11 December – 13 December 1964 | Installed spare two cubic foot demineralizer in additional shield water recirculation system to aid cleanup of shield water activity resulting from storage of spent core. PM-3A Operating Report Number 7; Malfunction-64-83. |
|---|---|
| 14 December – 16 December 1964 | Removed dummy source tube from new core and installed a PO-BE start-up source. Replaced spent fuel tank recirculating pump. Calibrated nuclear instrumentation for new core loading. |
| 17 December 1964 | Installed new core Type I, serial 1, in reactor pressure vessel. Completed installation and testing of McMurdo Station evacuation alert alarm. |
| 18 December – 19 December 1964 | Completed final phase of core loading. Finished construction of condenser underfloor storage area. Tested all plant safety valves. |
| 20 December – 26 December 1964 | Repaired control rod drive mechanism collets to meet diameter specifications and reassembled in reactor after refueling. Prepared and shipped old collet assembly to CONUS for evaluation. Performed control rod drive mechanism latching and 3/8 inch pickup procedures. |
| 27 December 1964 02 January 1965 | Packaged solid radioactive waste for shipment. Welded cracks in steam generator tank sump liner. Completed control rod actuator dimension checks and marked actuator wiring. Replaced steam generator and expansion tank pressure relief valves with CONUS tested and certified units. |
| 03 January – 07 January 1965 | Completed installation of electrical penetrations and painting of containment. Completed cold hydrostatic test of the primary system. Pressurized containment to 30 psig to aid in locating containment tank leaks. |
| 08 January – 15 January 1965 | Completed annual containment leak rate test IAW Special Test Number T-53 with satisfactory results. Initiated instrumentation calibration and tests. Completed primary sample system modification. |

8

| | |
|---|---|
| 16 January – 21 January 1965 | Completed instrumentation calibration and precritical checks. Completed calibration of the steam generator and pressurizer level control  Conducted five control rod drops. Control rod 1 cable found to be grounded. |
| 22 January – 25 January 1965 | Control rods failed to drive in "bank" or "three rod" modes of operation. Containment opened, control rod drive mechanisms inspected and found to have sustained water damage. Removed CRDM units and began drying five coil can cables, and three coil cans to remove moisture. PM-3A Operating Report Number 8; Malfunction-65-7. |
| 26 January – 28 January 1965 | Completed drying and venting of two coil cans to satisfactory condition. Two coil cans (serial numbers 202 and 203) could not be satisfactorily repaired and were prepared for shipment to CONUS for emergency repair. |
| 29 January – 30 January 1965 | Coil cans serial 202 and 203 backloaded to CONUS. BUDOCKS Inspection Team arrived for annual PM-3A safety and administration inspection. |
| 31 January – 01 February 1965 | In process of backloading radioactive waste and other material to CONUS. |
| 02 February – 04 February 1965 | Completed filling one high level waste cask with resin; second cask partially filled. PM-3A Operating Report Number 9; Annex I. |
| 05 February – 08 February 1965 | Completed backloading six CONEX boxes each containing twenty-three drums of radioactive solid waste and two high level casks containing resins. In process of completing new caterpillar diesel-generator building. |
| 09 February – 20 February 1965 | Plant in cold iron status awaiting arrival of repaired control rod drive mechanisms and new cables. |
| 21 February 1965 | Received CONUS repaired coil cans 202 and 203 with associated cables. PM-3A Operating Report Number 9; Annex II. |

9

| | |
|---|---|
| 22 February 1965 | Installed coil can housings and coil can cables on the reactor pressure vessel head. Installing position indicator cans and new cables. Completed caterpillar D-G building. |
| 23 February - 25 February 1965 | Completed installation of CRDM's latched and performed required control rod pick up tests. Completed cold control rod drop tests and primary system cold hydro test. Steam generator drained and flushed. Instrumentation and electrical precritical checks completed. Initiated radiochemistry sampling. In process of cleaning up crud and reducing primary coolant system activity level with demineralizer. |
| 26 February 1965 | Plant reached initial criticality on the new core at 1126 local. Plant shutdown at 1430 local to adjust limit switch. Plant critical at 1719 local. Plant shutdown at 1833 local. Plant critical at 2305 local. PM-3A Operating Report Number 9; Malfunction-65-16. |
| 27 February 1965 | Plant shutdown at 0457 local. Critical at 1833 local. Conducting shutdown margin tests. PM-3A Operating Report Number 9; Malfunction-65-17. |
| 28 February 1965 | Reactor scrammed during temperature coefficient testing when control rod number 5 dropped. Reactor scrammed twice on fast period during temperature coefficient testing. PM-3A Operating Report Number 10; Malfunction-65-18, 65-19, 65-20. |
| 01 March 1965 | Reactor critical for core physics testing. |
| 02 March 1965 | Reactor scrammed on fast period during temperature coefficient testing. PM-3A Operating Report Number 10; Malfunction-65-21. |
| 03 March 1965 | Reactor scrammed on low primary coolant pump power following completion of temperature coefficient tests. Temperature coefficient and six rod bank worth tests, six rod bank position vs temperature data, reactivity insertion rate, and radiochemical sampling performed. Plant shutdown due to loss of main steam pressure and steam generator level |

indication. PM-3A Operating Report Number 10; Malfunction-65-23.

| | |
|---|---|
| 04 March –<br>05 March 1965 | Plant shutdown. Cleaned steam generator level reference column. Replaced feedwater check valve internals. Nuclear instrumentation channel number 3 failed. PM-3A Operating Report Number 10; Malfunction-65-24. |
| 06 March 1965 | Repaired nuclear instrumentation intermediate range channel number 3 drawer and cables. Reactor scrammed during startup on fast period. PM-3A Operating Report Number 10; Malfunction-65-25. |
| 07 March 1965 | Plant placed in cold iron status awaiting CONUS evaluation of steam generator malfunction. |
| 08 March –<br>17 March 1965 | Plant in cold iron status. Completed primary system cold hydrostatic test. Completed wiring for scram circuit self test and display system modification. Disassembled and inspected steam generator level system datum column and calibrated steam generator level system. |
| 18 March 1965 | Calibrated feedwater flow control valve. Plant shutdown. Self test display drawer modification failed to test out satisfactorily, drawer wiring returned to original configuration at CONUS direction. Replaced detectors on nuclear instrumentation (intermediate range channel number 3 and power range channel number 7. |
| 19 March 1965 | Control rod number 5 would not drive properly during pre-startup tests. Reactor brought critical for continuation of core physics and primary system testing. PM-3A Operating Report Number 10; Malfunction-65-26. |
| 20 March 1965 | Reactor scrammed on fast period during zero power operation. PM-3A Operating Report Number 10; Malfunction-65-27. |

| | |
|---|---|
| 21 March –<br>22 March 1965 | Reactor critical. Following tests completed:<br>1. Temperature coefficient and six rod bank position vs temperature.<br>2. Pressurizer level calibration.<br>3. Steam generator level calibration.<br>4. Primary system instrument tests.<br>5. Pressurizer spray valve response. |
| 23 March 1965 | Reactor scrammed on instrumentation transient prior to warming up main steam line. Control rod number 5 would not drive during pre-startup tests. Reactor brought critical for secondary system shakedown tests. Reactor scrammed on loss of power from McMurdo Station Diesel plant with main turbine generator at synchronous speed and no load. PM-3A Operating Report Number 10; Malfunctions-65-28, 65-29, 65-30. |
| 24 March 1965 | Reactor brought critical for secondary system shakedown tests. Reactor scrammed by instrumentation switching transient. PM-3A Operating Report Number 10; Malfunction-65-31. |
| 25 March 1965 | Reactor brought critical for secondary system shakedown. Plant load picked up on main turbine generator. Turbine generator set vibration tests completed. |
| 26 March 1965 | Reactor critical, turbine generator on the line carrying plant load. Reactor scrammed on a transient while preparing to shut plant down. PM-3A Operating Report Number 10; Malfunction-65-32. |
| 27 March 1965 | Reactor shutdown. Automatic mode of feedwater control system malfunctioning. |
| 28 March 1965 | Reactor brought critical for nuclear instrumentation power range channel testing. Reactor shutdown to replace decay heat system check valve. Reactor brought critical and plant load picked up on main turbine generator. |
| 29 March 1965 | Reactor critical with plant load being carried on the main turbine generator. Assumed McMurdo Station |

electrical load. Continued secondary system shake-down tests.

| | |
|---|---|
| 30 March 1965 | Plant on the line carrying Mc Murdo Station load. Reactor scrammed on high power. PM-3A Operating Report Number 10; Malfunction-65-33. |
| 31 March 1965 | Assumed plant and McMurdo Station load for continuous power operations. Steam generator blowdown cooler partially blocked. |
| 01 April – 03 April 1965 | Plant on the line carrying McMurdo Station load. Checked calibration of steam and feedwater flow systems. Automatic mode of feedwater control system still malfunctioning. Main condenser fan 2B out of service. |
| 04 April – 27 April 1965 | Plant up for continuous power operations. During this period, the steam generator blowdown chemistry analysis was outside specified limits three times. Automatic mode of feedwater control system inoperative. PM-3A Operating Report Number 11; Malfunctions-65-34, 65-35, 65-36. |
| 28 April 1965 | Reactor scrammed on low primary coolant pressure when control rod number 6 slipped. Main generator was carrying plant load at time of scram. McMurdo Station load had been picked up by the McMurdo Station diesel plant to allow maintenance on the electrical distribution system by Public Works personnel. Fire broke out on top of Caterpillar diesel generator. PM-3A Operating Report Number 11; Malfunctions-65-37, 65-38. |
| 29 April – 30 April 1965 | Attempted to startup but had to shutdown due to inability to maintain primary system operating pressure. Replaced pressurizer heater element #13. Added high voltage monitors to nuclear instrumentation power range channels. Installed rod drop test modification to signal generator and logic drawer in control rod drive mechanism cabinet. PM-3A Operating Report Number 11; Malfunction-65-39. |

| | |
|---|---|
| 01 May 1965 | Reactor was brought critical and assumed plant load. Plant load was alternately dropped and picked up twice for operator training. |
| 02 May – 05 May 1965 | Plant up for power operations. |
| 06 May 1965 | Secondary system secured for maintenance on feed-water flow control valve bypass line (leak). Plant scrammed on fast period. A second reactor scram occurred during startup, turbine generator at synchronous speed, caused by operator error during sychronization with McMurdo Station Diesel Plant. A blown fuse in the synchronizing circuit produced false synchronizing information. PM-3A Operating Report Number 12; Malfunctions-65-40, 65-41. |
| 07 May – 21 May 1965 | Plant up for power operations. |
| 22 May 1965 | Plant up for power operations. Steam generator blow-down chemistry out of specified limits. PM-3A Operating Report Number 12; Malfunction-65-42. |
| 23 May 1965 | Plant up for power operations. Main condenser number one was frozen while attempting to place it into operation. PM-3A Operating Report Number 12; Malfunction-65-43. |
| 24 May – 09 June 1965 | Plant up for power operations. |
| 10 June 1965 | Plant up for power operations. Steam generator blowdown chemistry analysis out of specified limits. PM-3A Operating Report Number 13; Malfunction-65-44. |
| 11 June – 16 June 1965 | Plant up for power operations. |
| 17 June 1965 | Plant scrammed due to low primary pressure. Primary system hydrogen concentration out of specified limits. PM-3A Operating Report Number 13; Malfunctions-65-45, 65-46. |

14

| | |
|---|---|
| 18 June -<br>22 June 1965 | Plant shutdown for scheduled and unscheduled maintenance. Containment opened. Made cold hydrostatic test of primary system, steam generator and expansion tank. Inspected all containment instrumentation. Checked pressurizer heaters. Inspected steam generator level system. Completed containment electrical equipment megger tests. |
| 23 June -<br>28 June 1965 | Plant up for power operations. |
| 29 June 1965 | Plant up for power operations. Main condenser number one frozen while attempting to place into service. PM-3A Operating Report Number 13; Malfunction-65-47. |
| 30 June -<br>09 July 1965 | Plant up for power operations. |
| 10 July 1965 | Plant up for power operations. Nuclear instrumentation channel number 6 failed. PM-3A Operating Report Number 14; Malfunction-65-48. |
| 11 July 1965 | Plant up for power operations. Nuclear instrumentation channel number 6 failed. PM-3A Operating Report Number 14; Malfunction-65-49. |
| 12 July 1965 | Plant up for power operations. Containment cooler fan #1 tripped. Power to motor control center #2 was lost while attempting to restart containment cooler fan #1. PM-3A Operating Report Number 14; Malfunction 65-60. |
| 13 July -<br>24 July 1965 | Plant up for power operations. |
| 25 July 1965 | Plant up for power operations. Feedwater system oxygen concentration out of specified limits. PM-3A Operating Report Number 14; Malfunction-65-51. |
| 26 July -<br>27 July 1965 | Plant up for power operations. |

| | |
|---|---|
| 28 July 1965 | Plant up for power operations. Two fuses blew in control rod actuator 12 volt power supply and the lift-pulldown power supply. PM-3A Operating Report Number 14; Malfunction-65-52. |
| 29 July - 07 August 1965 | Plant up for power operations. |
| 08 August 1965 | Reactor scrammed due to momentary interruption of 65 volt power supply output in control console. PM-3A Operating Report Number 15; Malfunction-65-53. |
| 09 August - 11 August 1965 | Plant shutdown for scheduled and unscheduled maintenance. |
| 12 August 1965 | Reactor control rod number 1 was sluggish while attempting to achieve criticality. Reactor critical for primary system hot leak rate and shutdown margin tests. PM-3A Operating Report Number 15; Malfunction-65-54. |
| 13 August 1965 | Reactor shutdown to reset control rod limit switch and further evaluate reactor control rod number 1 withdrawal response while in a hot shutdown condition. Reactor brought critical. |
| 14 August - 16 August 1965 | Plant up for power operations. |
| 17 August 1965 | Plant up for power operations. Nuclear instrumentation power channel number 6 failed. PM-3A Operating Report Number 15; Malfunction-65-55. |
| 18 August 1965 | Plant up for power operations. McMurdo Station load was dropped for approximately four hours to allow station Public Works personnel to perform maintenance on the power distribution system. |
| 19 August - 20 August 1965 | Plant up for power operations. |
| 21 August 1965 | Plant scrammed due to switching transient in turbine generator speed governor switch. Reactor brought |

16

critical. Shutdown secondary system due to inability to obtain over ten inches of mercury vacuum. PM-3A Operating Report Number 15; Malfunctions-65-56, 65-57.

| | |
|---|---|
| 22 August 1965 | Plant up for power operations. Reactor control rod drive mechanism malfunctioned. Plant manually scrammed due to loss of main condenser vacuum. Reactor brought critical. PM-3A Operating Report Number 15; Malfunctions-65-58, 65-59. |
| 23 August – 26 August 1965 | Plant up for power operations. |
| 27 August 1965 | Plant scrammed due to momentary short circuit of 24 volt DC power supply in control console. Reactor brought critical. PM-3A Operating Report Number 15; Malfunction-65-60. |
| 28 August – 02 September 1965 | Plant up for power operations. |
| 03 September 1965 | Plant up for power operations. Steam generator blowdown activity out of specified limits. PM-3A Operating Report Number 15; Malfunction-65-61. |
| 04 September – 05 September 1965 | Plant up for power operations. |
| 06 September 1965 | Plant scrammed due to apparent electrical transient. Reactor brought critical and plant assumed McMurdo Station load. PM-3A Operating Report Number 16; Malfunction-65-62. |
| 07 September 1965 | Plant scrammed on 120% power. Reactor brought critical, preparing to place secondary system in operation, when plant scrammed due to fast reactor period. PM-3A Operating Report Number 16; Malfunctions-65-63, 65-64. |
| 08 September 1965 | Plant scrammed while withdrawing control rods to achieve criticality. Reactor brought critical, plant assumed McMurdo Station load. PM-3A Operating Report Number 16; Malfunction-65-65. |

| | |
|---|---|
| 09 September 1965 | Plant up for power operations. Lost time accident occurred. PM-3A Operating Report Number 16; Malfunction-65-66. |
| 10 September – 11 September 1965 | Plant up for power operations. |
| 12 September 1965 | Plant scrammed due to apparent electrical transient. Brought reactor critical. PM-3A Operating Report Number 16; Malfunction-65-67. |
| 13 September – 21 September 1965 | Plant up for power operations. |
| 22 September 1965 | Plant scrammed due to a short circuit in the primary system leak detector. Steam generator blowdown chemistry analyses out of normal specified limits. PM-3A Operating Report Number 16; Malfunctions-65-68, 65-69. |
| 23 September – 24 September 1965 | Plant shutdown. Containment opened. Performed numerous maintenance items and completed required 2500 hour tests. |
| 25 September 1965 | Closed containment, reactor brought critical, and assumed McMurdo Station load. |
| 26 September 1965 | Plant up for power operation. Nuclear Instrumentation power range channel 6 failed. PM-3A Operating Report Number 16; Malfunction-65-70. |
| 27 September 1965 | Reactor scrammed due to accidental main steam stop valve closure. Reactor brought critical and plant assumed McMurdo Station load. PM-3A Operating Report Number 16; Malfunction-65-71. |
| 28 September – 29 September 1965 | Plant up for power operations. |
| 30 September 1965 | Plant up for power operations. Nuclear Instrumentation power range channel 6 failed. PM-3A Operating Report Number 16; Malfunction-65-72. |

18

01 October –
07 October 1965

Plant up for power operations

08 October –
09 October 1965

Plant scrammed due to unknown cause followed by
three scrams attributed to switching transients and
one training scram. PM-3A Operating Report Number
17; Malfunction-65-73.

10 October –
15 October 1965

Reactor brought critical and plant assumed McMurdo
Station load.

16 October 1965

Plant scrammed manually when control rod number 2 droppe
Reactor brought critical and assumed plant load. Be-
gan cycling of plant for Crew V training. PM-3A
Operating Report Number 17; Malfunction-65-74.

17 October 1965

Plant scrammed for training. Reactor brought
critical and assumed plant load. Continued plant
cycling for training.

18 October 1965

Plant cycling for training with one scheduled scram.
Reactor brought critical and assumed McMurdo
Station load.

19 October –
22 October 1965

Plant up for power operations.

23 October 1965

Commenced cycling plant for training, scheduled
scram and shutdown for maintenance. Commenced
containment purge.

24 October –
30 October 1965

Plant down for maintenance.

31 October 1965

Reactor critical for core physics testing. Reactor
scrammed on an indicated 30 second period. Reactor
brought critical and scrammed on a short period.
Reactor brought critical. PM-3A Operating Report
Number 18; Malfunctions-65-76, 65-77, 65-78.

01 November 1965

Reactor critical for core physics testing. Reactor
scrammed due to electrical transient. Reactor
brought critical and them scrammed for scheduled

maintenance. PM-3A Operating Report Number 18; Malfunction-65-79.

| | |
|---|---|
| 02 November - 15 November 1965 | PM-3A down for scheduled maintenance. |
| 16 November 1965 | Reactor critical in preparation for power operations. Reactor scrammed due to electrical transient. Reactor brought critical and again scrammed due to short period on nuclear instrumentation channels numbers 3 and 4    PM-3A Operating Report Number 18; Malfunctions-65-81, 65-82. |
| 17 November 1965 | Reactor brought critical. Scrammed on electrical transient in 4160V distribution system. Reactor brought critical, assumed McMurdo Station load. PM-3A Operating Report Number 18; Malfunction-65-83. |
| 18 November 1965 | PM-3A up for power operations. |
| 19 November 1965 | Dropped McMurdo Station and PM-3A load. Repaired leak in feedwater system line. Assumed PM-3A and McMurdo Station heater bank load. |
| 20 November 1965 | Assumed McMurdo Station industrial load. |
| 21 November - 31 November 1965 | Plant up for power operations. |
| 01 December 1965 | PM-3A on the line for power operations. Secondary system chemistry out of normal specifications. PM-3A Operating Report Number 19; Malfunction-65-85. |
| 02 December 1965 | PM-3A on the line for power operations. |
| 03 December 1965 | PM-3A on the line for power operations. Released liquid waste in excess of prescribed limits. PM-3A Operating Report Number 19; Malfunction-65-86. |
| 04 December - 05 December 1965 | PM-3A on the line for power operations. |

| 06 December 1965 | Dropped McMurdo Station and plant load. Shutdown reactor to obtain information on excessive leak rate. Purging containment tanks. |
|---|---|
| 07 December - 08 December 1965 | PM-3A shutdown for maintenance. |
| 09 December 1965 | PM-3A shutdown for maintenance. Radiation exposure of personnel in excess of 300 MREM/WK. PM-3A Operating Report Number 19; Malfunction-65-87. |
| 10 December 1965 | PM-3A shutdown for maintenance. |
| 11 December 1965 | Reactor brought critical. |
| 12 December 1965 | Assumed McMurdo Station load. |
| 13 December - 26 December 1965 | PM-3A on the line for power operations. |
| 27 December 1965 | PM-3A on the line for power operations. Received and stored PM-3A core III and 55 curie Po-Be startup source. |
| 28 December 1965 - 07 January 1966 | PM-3A on the line for power operations. |
| 08 January 1966 | PM-3A up for power operations. Primary coolant chemistry outside normal operating limits. PM-3A Operating Report Number 20; Malfunction-66-1. |
| 09 January - 10 January 1966 | PM-3A up for power operations. |
| 11 January - 23 January 1966 | PM-3A shutdown to remove spent core from storage. |
| 24 January 1966 | Reactor critical. PM-3A assumed McMurdo Station load. |
| 25 January 1966 | PM-3A up for power operations. Primary coolant chemistry outside normal operating limits. PM-3A Operating Report Number 20; Malfunction-66-3. |

| | |
|---|---|
| 26 January 1966 | PM-3A shutdown for inability to maintain primary coolant chemistry within normal operating limits. |
| 27 January –<br>29 January 1966 | PM-3A shutdown. Purging containment in preparation for changing primary demineralizer resins. |
| 30 January –<br>05 February 1966 | PM-3A delayed in startup due to a malfunction in the Radioactive Waste Disposal System (RWDS). PM-3A Operating Report Number 21; Malfunction-66-5. |
| 06 February 1966 | Reactor critical, had to shut down due to bad power supply in the rod actuator cabinet. PM-3A Operating Report Number 21; Malfunction-66-8. |
| 07 February 1966 | Reactor critical. Could not drive rods past eleven inches. PM-3A Operating Report Number 21; Malfunction-66-9. |
| 08 February –<br>11 February 1966 | PM-3A shutdown for Control Rod Drive Mechanism (CRDM) repairs. |
| 12 February 1966 | Reactor critical. Assumed McMurdo Station load. |
| 13 February 1966 | PM-3A on the line for power operations. |
| 14 February 1966 | Reactor scrammed and then brought critical. Assumed McMurdo Station load. PM-3A Operating Report Number 21; Malfunction-66-6. |
| 15 February –<br>18 February 1966 | PM-3A on the line for power operations. |
| 19 February –<br>20 February 1966 | PM-3A on the line for power operations. Transported approximately 20,000 lbs of steam to water distillation plant for testing and distillation of approximately 4,000 gallons of fresh water. |
| 21 February –<br>10 April 1966 | PM-3A on the line for power operations. |
| 11 April 1966 | PM-3A on the line for power operations. Bistable number 12 tripped at approximately 98% power. PM-3A Operating Report Number 23; Malfunction 66-11. |

| | |
|---|---|
| 12 April - 20 April 1966 | PM-3A on the line for power operations. |
| 21 April 1966 | PM-3A on the line for power operations. Reactor coolant pump lower bearing temperature indication became erratic. PM-3A Operating Report Number 23; Malfunction-66-12. |
| 22 April - 28 April 1966 | PM-3A on the line for power operations. |
| 29 April 1966 | PM-3A on the line for power operations. Loss of vacuum and control of main condenser Number three, necessitated paralleling the PM-3A with McMurdo Station Diesel Plant for a short period of time. PM-3A Operating Report Number 23; Malfunction 66-13. |
| 30 April - 08 May 1966 | PM-3A on the line for power operations. |
| 09 May 1966 | PM-3A on the line for power operations. Unscheduled reactor scram at 0220. PM-3A Operating Report Number 24; Malfunction-66-14. |
| 10 May - 15 May 1966 | PM-3A remained shutdown for scheduled and unscheduled maintenance. |
| 16 May 1966 | Reactor brought critical for core physics testing. Unscheduled reactor scram. PM-3A Operating Report Number 24, Malfunction-66-15. |
| 17 May 1966 | Reactor brought critical. Reactor scrammed. Reactor brought critical. PM-3A Operating Report Number 24; Malfunction-66-16. |
| 18 May 1966 | Reactor critical. PM-3A assumed McMurdo Station load. |
| 19 May 1966 | PM-3A on the line for power operations. |
| 20 May 1966 | PM-3A on the line for power operations. Unscheduled reactor scram at 1120; reactor brought critical. |

23

Reactor scrammed. Reactor brought critical.
PM-3A Operating Report Number 24; Malfunctions-
66-17, 66-18.

21 May 1966          Reactor critical. PM-3A assumed McMurdo Station
load.

22 May 1966          PM-3A on the line for power operations. Failure of
main condenser fan 1B. PM-3A Operating Report
Number 24; Malfunction-66-19.

23 May -
16 August 1966        PM-3A on the line for power operations.

17 August 1966       PM-3A on the line for power operations. PM-3A
switched to morpholine control. PM-3A Operating
Report Number 27; Section IV, Item 4.

18 August -
07 October 1966      PM-3A on the line for power operations.

08 October 1966      PM-3A on the line for power operations. At 0101,
the PM-3A surpassed the record for the longest
continuous power run for nuclear power plants operated
by military personnel.

09 October 1966      PM-3A on the line for power operations. Reactor
manually scrammed at 1155. Reactor brought cri-
tical. PM-3A Operating Report Number 29; Mal-
function-66-20.

10 October 1966      PM-3A assumed McMurdo Station load.

11 October -
14 October 1966      PM-3A on the line for power operations.

15 October 1966      PM-3A on the line for power operations. Reactor
scrammed at 1155. Reactor was brought critical,
plant load assumed. Reactor manually scrammed
several times for training of new crew. PM-3A
Operating Report Number 29; Malfunction-66-21.

16 October 1966      Assumed McMurdo Station load for power operations.

24

| 17 October –<br>21 October 1966 | PM-3A on the line for power operations. |
|---|---|
| 22 October 1966 | PM-3A dropped McMurdo Station load. Cycling of plant for crew training. |
| 23 October 1966 | Plant being cycled for crew training. Reactor scrammed at 0855. PM-3A Operating Report Number 29; Malfunction-66-22. |
| 24 October –<br>06 November 1966 | PM-3A shutdown for annual maintenance. |
| 07 November 1966 | Reactor brought critical and then scrammed at 2113. Reactor brought critical. PM-3A Operating Report Number 30; Malfunction-66-23. |
| 08 November 1966 | Reactor scrammed at 1400. PM-3A Operating Report Number 30; Malfunction-66-24. |
| 09 November –<br>10 November 1966 | PM-3A shutdown to correct misalignment of CRDM. |
| 11 November 1966 | Reactor brought critical at 0536 for core physics testing. |
| 12 November 1966 | PM-3A continued core physics testing. A fire was discovered and extinguished in under condenser storage area. PM-3A Operating Report Number 30; Malfunction-66-25. |
| 13 November 1966 | Reactor scrammed at 0700 on noise transient. Reactor brought critical at 1042 and scrammed at 2320 on noise transient. PM-3A Operating Report Number 30; Malfunctions-66-26, 66-27. |

| | |
|---|---|
| 14 November 1966 | PM-3A performing test procedure -- RC-2, control rod drop time. Power lost to channel number 5. Rod no. 1 appeared unlatched. PM-3A Operating Report Number 30; Malfunctions-66-28, 66-29. |
| 15 November - 27 November 1966 | PM-3A shutdown, containment open in preparation for replacement of control rod number 1. PM-3A special operating report for broken control rod replacement prepared 28 February 1967. |
| 28 November 1966 | Reactor brought critical. Reactor scrammed at 2259. PM-3A Operating Report Number 31; Malfunction-66-30. |
| 29 November 1966 | Reactor brought critical; scrammed at 0123. Reactor brought critical, assumed plant load at 2056, dropped plant load at 2220. PM-3A Operating Report Number 31; Malfunctions-66-31, 66-32. |
| 30 November 1966 | PM-3A assumed McMurdo Station load at 0525, dropped McMurdo Station load at 0610, assumed McMurdo Station load at 1350. |
| 01 December 1966 | Reactor scrammed at 1050. Reactor brought critical at 2230. PM-3A Operating Report Number 31; Malfunctions-66-33, 66-34. |
| 02 December 1966 | Assumed McMurdo Station load at 0444. |
| 03 December 1966 - 01 February 1967 | PM-3A on the line for power operations. |
| 02 February 1967 | Reactor scrammed at 1627. PM-3A Operating Report Number 33; Malfunctions-67-2, 67-3. |
| 03 February - 08 February 1967 | PM-3A shutdown for unscheduled maintenance |
| 09 February 1967 | Reactor scrammed at 2034 while being brought critical. Reactor brought critical. PM-3A Operating Report Number 33; Malfunction--67-4. |
| 10 February 1967 | Reactor scrammed at 0735. PM-3A Operating Report Number 33; Malfunction-67-5. |

26

| 11 February 1967 | Reactor brought critical at 2041. |
| --- | --- |
| 12 February 1967 | PM-3A assumed McMurdo Station load at 0624. Reactor scrammed at 1345. Reactor brought critical at 1643. PM-3A assumed plant load at 2035 and assumed Station load at 2155. PM-3A Operating Report Number 33; Malfunction-67-7, 67-8. |
| 13 February - 08 April 1967 | PM-3A on the line for power operations. |
| 09 April 1967 | Reactor scrammed at 0943. PM-3A Operating Report Number 35; Malfunction-67-9. |
| 10 April - 14 April 1967 | PM-3A shutdown for scheduled and unscheduled maintenance. |
| 15 April 1967 | Reactor brought critical at 0836. Reactor scrammed at 1123. Reactor brought critical at 1248. Reactor scrammed at 1417 during planned shutdown. PM-3A Operating Report Number 35; Malfunctions-67-10, 67-11, 67-12. |
| 16 April 1967 | Reactor critical at 1020, PM-3A assumed McMurdo Station load at 1735. |
| 17 April - 14 September 1967 | PM-3A on the line for power operations. |
| 15 September 1967 | Reactor scram. -- NI channel number 1 failed. PM-3A Operating Report Number 40; Malfunctions-67-18, 67-19. |
| 16 September - 20 September 1967 | PM-3A shutdown for testing and maintenance. |
| 21 September 1967 | Reactor brought critical. |
| 22 September 1967 | Two reactor scrams occurred due to faulty main steam stop valve actuator. PM-3A Operating Report Number 40; Malfunctions-67-20, 67-21. |
| 23 September 1967 | Reactor critical. PM-3A assumed McMurdo Station load. |

| | |
|---|---|
| 24 September – 25 September 1967 | PM-3A on the line for power operations. |
| 26 September 1967 | Reactor scrammed. PM-3A Operating Report Number 40; Malfunction-67-22. Reactor brought critical. |
| 27 September 1967 | Reactor scrammed. Reactor brought critical. PM-3A assumed McMurdo Station load. PM-3A Operating Report Number 40; Malfunction-67-23. |
| 28 September – 08 October 1967 | PM-3A on the line for power operations. |
| 09 October 1967 | Began cycling plant for relief crew training. |
| 12 October 1967 | Plant shutdown for core change. PM-3A Operating Report Number 41; Annex III. |
| 13 October 1967 | Containment opened. Control rod drive mechanism removal started. |
| 14 October 1967 | Completed control rod drive mechanism removal. |
| 16 October 1967 | Pressure vessel head removal completed. The old core was placed in the spent core tank. |
| 18 October 1967 | The new core was installed in the pressure vessel. |
| 20 October 1967 | Completed replacement of pressure vessel head. |
| 21 October 1967 | Completed installation of control rod drive mechanism. |
| 22 October 1967 | All control rod latching completed. |
| 23 October 1967 | Initial criticality achieved with new type II core, and initial startup testing began. Reactor scrammed on short period. Control rod number 5 dropped from critical position. Reactor brought critical. PM-3A Operating Report Number 41; Malfunctions-67-24, 67-25. |
| 24 October 1967 | Reactor scrammed on short period. Reactor brought critical. PM-3A Operating Report Number 41; Malfunction-67-26. |

| | |
|---|---|
| 25 October 1967 | Reactor scrammed when the McMurdo Station Diesel Plant opened the PM-3A tieline breaker. Reactor brought critical. Reactor scrammed when the McMurdo Station diesel generator number 2 tripped off the line. PM-3A Operating Report Number 41; Malfunctions-67-27, 67-28. |
| 26 October 1967 | Reactor brought critical. Reactor scrammed when the Caterpillar diesel generator tripped off the line. PM-3A Operating Report Number 41; Malfunction-67-29. |
| 27 October 1967 | Reactor critical for testing. |
| 28 October 1967 | Pressurizer level system out of calibration. Reactor scrammed on high primary system pressure. Sluggishness in control rod drive system. PM-3A Operating Report Number 41; Malfunctions-67-30, 67-31, 67-32. |
| 29 October – 01 November 1967 | Plant shutdown. |
| 02 November 1967 | Reactor brought critical, shutdown, and brought critical. |
| 03 November 1967 | Reactor twice shutdown and brought critical. PM-3A assumed plant load. PM-3A Operating Report Number 42; Malfunction-67-33. |
| 04 November 1967 | PM-3A assumed McMurdo Station load. |
| 05 November – 29 November 1967 | PM-3A on the line for power operations. |
| 30 November 1967 | PM-3A operated isolated while Public Works Department inspected the switching station breakers. |
| 30 November – 01 December 1967 | PM-3A on the line for power operations. |
| 02 December 1967 | Reactor scrammed twice. PM-3A Operating Report Number 43; Malfunctions-67-34, 67-35. |

| | |
|---|---|
| 03 December – 27 December 1967 | PM-3A on the line for power operations. |
| 28 December 1967 | Reactor scrammed. Reactor critical -- picked up McMurdo Station load. PM-3A Operating Report Number 43; Malfunction-67-36. |
| 29 December 1967 – 01 January 1968 | PM-3A on the line for power operations. |
| 02 January – 17 January 1968 | PM-3A shutdown for scheduled turbine inspection and scheduled maintenance. |
| 18 January 1968 | Reactor brought critical and shutdown. PM-3A Operating Report Number 44; Malfunction-68-1. |
| 19 January 1968 | Reactor brought critical. PM-3A assumed McMurdo Station load. |
| 20 January – 29 January 1968 | PM-3A on the line for power operations. |
| 30 January 1968 | Reactor scrammed. PM-3A Operating Report Number 45; Malfunction-68-2. |
| 31 January 1968 | Reactor brought critical. PM-3A assumed McMurdo Station load. |
| 01 February – 10 February 1968 | PM-3A on the line for power operations. |
| 11 February – 14 February 1968 | PM-3A shutdown to performed scheduled maintenance. |
| 15 February 1968 | Reactor brought critical. PM-3A assumed McMurdo Station load. |
| 16 March – 25 March 1968 | PM-3A on the line for power operations. |
| 26 March – 30 March 1968 | PM-3A shutdown to perform unscheduled maintenance. |

| | |
|---|---|
| 31 March 1968 | Reactor brought critical. PM-3A assumed McMurdo Station load. |
| 01 April 1968 | Secondary plant shutdown to perform unscheduled maintenance. Reactor brought critical. PM-3A assumed McMurdo Station load. PM-3A Operating Report Number 47; Malfunction-68-8. |
| 02 April – 20 June 1968 | PM-3A on the line for power operations. |
| 21 June 1968 | Reactor scrammed. PM-3A Operating Report Number 49; Malfunctions-68-10, 68-11. |
| 22 June 1968 | Reactor brought critical. PM-3A assumed McMurdo Station load. |
| 23 June – 19 July 1968 | PM-3A on the line for power operations. |
| 19 July 1968 | Reactor in the process of shutdown. Reactor scram. PM-3A Operating Report Number 50; Malfunctions-68-12, 68-13. |
| 20 July – 23 July 1968 | Plant shutdown. |
| 24 July 1968 | Reactor brought critical. PM-3A assumed McMurdo Station load. |
| 25 July – 26 July 1968 | PM-3A on the line for power operations. |
| 26 July 1968 | Reactor scrammed. Reactor brought critical. PM-3A assumed McMurdo Station load. PM-3A Operating Report Number 50; Malfunctions-68-14, 68-15. |
| 26 July – 12 September 1968 | PM-3A on the line for power operations. |
| 13 September 1968 | Reactor scrammed. PM-3A Operating Report Number 50; Malfunction-68-19. |

| | |
|---|---|
| 14 September – 22 September 1968 | Plant shutdown. |
| 23 September 1968 | Reactor critical. PM-3A assumed McMurdo Station load. |
| 24 September – 25 September 1968 | PM-3A on the line for power operations. |
| 26 September 1968 | Reactor shutdown. Reactor brought critical. PM-3A assumed McMurdo Station load. PM-3A Operating Report Number 50; Malfunction-68-22. |
| 27 September – 15 October 1968 | PM-3A on the line for power operations. |
| 15 October 1968 | The site tieline breaker was manually opened to reduce KVAR load when the McMurdo Station Diesel plant paralleled with voltage too low to assume any reactive electrical load. |
| 15 October – 18 October 1968 | PM-3A on the line for power operations. |
| 18 October – 19 October 1968 | The secondary system was cycled eight times for replacement crew training by transferring the electrical load between the PM-3A and the McMurdo Station Diesel Plant. |
| 19 October – 21 October 1968 | PM-3A shutdown for maintenance. |
| 21 October – 23 October 1968 | The primary system (reactor) was cycled six times for replacement crew training. |
| 23 October – 24 October 1968 | Reactor critical, performing core physics tests. |
| 24 October – 20 November 1968 | PM-3A on the line for power operations. |
| 20 November 1968 | Reactor scrammed. PM-3A Operating Report Number 51; Malfunctions-68-27, 68-28. |

32

| | |
|---|---|
| 21 November 1968 | Reactor critical. Reactor scram. PM-3A Operating Report Number 51; Malfunctions-68-29, 68-30. |
| 21 November – 24 November 1968 | Plant shutdown. |
| 24 November 1968 | Reactor critical. PM-3A assumed McMurdo Station load. |
| 25 November – 06 December 1968 | PM-3A on the line for power operations. |
| 06 December 1968 | Reactor scrammed. Reactor brought critical. PM-3A assumed McMurdo Station load. PM-3A Operating Report Number 51; Malfunction-68-32. |
| 06 December – 10 December 1968 | PM-3A on the line for power operations. |
| 10 December 1968 | Secondary system shutdown to repair steam leak. PM-3A assumed McMurdo Station load. |
| 10 December 1968 – 01 January 1969 | PM-3A on the line for power operations. |
| 01 January 1969 – 02 January 1969 | Reactor scrammed. Reactor brought critical. PM-3A assumed McMurdo Station load. PM-3A Operating Report Number 51; Malfunctions-69-1, 69-2. |
| 02 January – 15 January 1969 | PM-3A on the line for power operations. |
| 15 January – 21 February 1969 | PM-3A shutdown for annual maintenance. PM-3A Test Procedure PS-1 (containment leak rate) was performed during this shutdown. |
| 22 February 1969 | Reactor brought critical. PM-3A assumed McMurdo Station load. |
| 23 February 1969 | PM-3A on the line supplying PM-3A (plant), Maintenance and Supply building, and the Water Distillation Plant with electrical power while maintenance was |

33

being performed on the McMurdo Station electrical distribution system. PM-3A assumed McMurdo Station load.

23 February – 07 March 1969

PM-3A on the line for power operations.

07 March – 08 March 1969

PM-3A shutdown. PM-3A Operating Report Number 52; Malfunction-69-9.

08 March 1969

Reactor brought critical.

09 March 1969

PM-3A assumed McMurdo Station load.

09 March – 13 March 1969

PM-3A on the line for power operations.

13 March 1969

PM-3A Secondary System shutdown. PM-3A assumed McMurdo Station load. PM-3A Operating Report Number 52; Malfunction-69-12.

13 March – 23 March 1969

PM-3A on the line for power operations.

23 March 1969

Reactor scram. Reactor brought critical. PM-3A assumed McMurdo Station load. PM-3A Operating Report Number 52; Malfunction-69-15.

23 March – 02 June 1969

PM-3A on the line for power operations.

02 June – 04 June 1969

PM-3A shutdown. PM-3A Operating Report Number 53; Malfunctions-69-16.

05 June 1969

Reactor brought critical.

06 June 1969

PM-3A assumed McMurdo Station load.

06 June – 14 June 1969

PM-3A on the line for power operations.

14 June – 15 June 1969

PM-3A shutdown. PM-3A Operating Report Number 53; Malfunction-69-20.

| | |
|---|---|
| 16 June 1969 | Reactor crtical. PM-3A assumed McMurdo Station load. |
| 16 June – 21 July 1969 | PM-3A on the line for power operations. |
| 21 July 1969 | Reactor scrammed. Reactor brought critical. PM-3A assumed McMurdo Station load. PM-3A Operating Report Number 54; Malfunction-69-23. |
| 21 July – 02 August 1969 | PM-3A on the line for power operations. |
| 02 August – 05 August 1969 | PM-3A shutdown. PM-3A Operating Report Number 54; Malfunction-69-25. |
| 06 August 1969 | Reactor critical. |
| 07 August 1969 | PM-3A assumed McMurdo Station load. |
| 07 August – 09 August 1969 | PM-3A on the line for power operations. |
| 09 August – 10 August 1969 | PM-3A shutdown. PM-3A Operating Report Number 54; Malfunction-69-30. |
| 11 August 1969 | Reactor critical. |
| 12 August 1969 | PM-3A assumed McMurdo Station load. |
| 12 August – 22 October 1969 | PM-3A on the line for power operations. |
| 22 October 1969 | Reactor scrammed. PM-3A Operating Report Number 55; Malfunction-69-33. |
| 22 October – 26 October 1969 | PM-3A shutdown. |
| 26 October – 27 October 1969 | The Primary System was being cycled for replacement crew training. |
| 27 October – 28 October 1969 | The Secondary System was being cycled for replacement crew training. |

35

| | |
|---|---|
| 28 October 1969 | Reactor manually scrammed. Reactor brought critical. PM-3A Operating Report Number 55; Malfunction-69-37. |
| 29 October 1969 | PM-3A assumed McMurdo Station load. |
| 29 October – 05 November 1969 | PM-3A on the line for power operations. |
| 05 November – 06 November 1969 | Secondary System shutdown. PM-3A assumed McMurdo Station load. PM-3A Operating Report Number 55; Malfunction-69-39. |
| 06 November – 26 November 1969 | PM-3A on the line for power operations. |
| 26 November 1969 | The site tieline breaker was opened for CBU 201 to facilitate relocation of switching station number one (Work Project M-50). |
| 25 November – 05 December 1969 | PM-3A on the line for power operations. |
| 05 December 1969 | Reactor shutdown for NAVFAC inspection. Reactor brought critical. |
| 06 December 1969 | PM-3A assumed McMurdo Station load. |
| 07 December – 08 December 1969 | The site tieline breaker was opened in order that CBU 201 could continue Work Project M-50. |
| 08 December 1969 | PM-3A assumed McMurdo Station load. |
| 08 December – 17 December 1969 | PM-3A on the line for power operations. |
| 17 December 1969 – 27 January 1970 | Reactor shutdown. It was decided to begin the annual maintenance shutdown. PM-3A Operating Report Number 55; Malfunctions-69-44, 69-46. |
| 27 January 1970 | Reactor critical for startup testing. |
| 28 January 1970 | PM-3A assumed plant load. |

36

| | |
|---|---|
| 29 January 1970 | PM-3A assumed McMurdo Station load. |
| 29 January – 31 January 1970 | PM-3A on the line for power operations. |
| 31 January 1970 | Reactor scrammed. PM-3A Operating Report Number 56; Malfunction-70-11. |
| 31 January – 09 February 1970 | PM-3A shutdown. |
| 10 February 1970 | Reactor critical. PM-3A assumed McMurdo Station load. |
| 10 February – 04 April 1970 | PM-3A on the line for power operations. |
| 04 April 1970 | Reactor scrammed. PM-3A Operating Report 56; Malfunction-70-23. |
| 05 April – 08 April 1970 | PM-3A shutdown. |
| 09 April 1970 | Reactor brought critical |
| 10 April 1970 | PM-3A assumed McMurdo Station load. |
| 10 April – 07 May 1970 | PM-3A on the line for power operations |
| 07 May 1970 | Main turbine-generator shutdown. PM-3A assumed McMurdo Station load. PM-3A Operating Report Number 57; Malfunction-70-27. |
| 07 May – 20 June 1970 | PM-3A on the line for power operations . |
| 21 June – 08 July 1970 | PM-3A shutdown for refueling with core type IV. PM-3A Operating Report Number 58; Appendix D. |
| 08 July – 15 July 1970 | Pre-startup testing and core physics testing of core physics testing of core type IV. |
| 15 July – 18 July 1970 | PM-3A on the line for power operations. |

| 18 July 1970 | Secondary System shutdown for core physics testing. PM-3A assumed McMurdo Station load. |
| --- | --- |
| 18 July – 17 September 1970 | PM-3A on the line for power operations. |
| 17 September 1970 | Reactor scrammed. Reactor brought critical. PM-3A assumed McMurdo Station load. PM-3A Operating Report Number 58; Malfunction-70-42. |
| 17 September – 02 October 1970 | PM-3A on the line for power operations. |
| 02 October 1970 | Plant shutdown. PM-3A Operating Report Number 58; Malfunction-70-47. |
| 03 October 1970 | Reactor brought critical. |
| 04 October 1970 | PM-3A assumed McMurdo Station load. Reactor scrammed. Reactor critical. PM-3A assumed Mc-Murdo Station load. PM-3A Operating Report Number 59; Malfunction-70-48. |
| 04 October – 05 October 1970 | PM-3A on the line for power operations. |
| 05 October – 06 October 1970 | Secondary plant system shutdown. PM-3A assumed McMurdo Station load. PM-3A Operating Report Number 59; Malfunction-70-49. |
| 06 October – 19 October 1970 | PM-3A on the line for power operations. |
| 19 October 1970 | Reactor shutdown. Reactor critical. PM-3A assumed McMurdo Station load. PM-3A Operating Report Number 59; Malfunction-70-50. |
| 19 October – 22 October 1970 | PM-3A on the line for power operations. |
| 26 October 1970 | The Primary System was being cycled for replacement crew training. |

| | |
|---|---|
| 26 October – 28 October 1970 | Delay in startup. Reactor critical. PM-3A assumed McMurdo Station load. PM-3A Operating Report Number 59; Malfunction-70-51. |
| 28 October – 29 October 1970 | PM-3A on the line for power operations and load transfer training. |
| 29 October 1970 | Reactor scrammed. Reactor critical. PM-3A Operating Report Number 59; Malfunction-70-52. |
| 29 October – 30 October 1970 | Secondary System cycling for training. |
| 30 October – 01 November 1970 | PM-3A on the line for power operations. |
| 01 November – 03 November 1970 | Reactor shutdown. Reactor critical. PM-3A Operating Report Number 59; Malfunction-70-53. |
| 03 November 1970 | Secondary System cycling for training. |
| 03 November – 10 November 1970 | PM-3A on the line for power operations. |
| 10 November 1970 | Reactor scram. PM-3A Operating Report Number 59; Malfunction-70-56. |
| 11 November 1970 | PM-3A assumed McMurdo Station load. |
| 11 November – 25 November 1970 | PM-3A on the line for power operations. |
| 25 November 1970 | Reactor scram. Reactor critical. PM-3A assumed McMurdo Station load. PM-3A Operating Report Number 59; Malfunction-70-58. |
| 25 November – 28 November 1970 | PM-3A on the line for power operations. |
| 28 November – 29 November 1970 | Site tie breaker opened. PM-3A Operating Report Number 59; Malfunction-70-60. |

| 29 November 1970 | PM-3A on the line for power operations. Reactor scram. PM-3A Operating Report Number 59; Malfunction-70-61. |
| --- | --- |
| 29 November – 04 December 1970 | Reactor critical -- delay in startup. PM-3A Operating Report Number 59; Malfunction-70-62. |
| 04 December – 05 December 1970 | Reactor shutdown. PM-3A Operating Report Number 59; Malfunction-70-63. |
| 05 December 1970 02 January 1971 | Reactor shutdown. At this time, it was decided to to remain shutdown for annual maintenance repair and overhaul. PM-3A Operating Report Number 59; Malfunction-70-65. |
| 03 January – 19 January 1971 | PM-3A shutdown for annual maintenance. |
| 19 January – 21 January 1971 | Reactor critical for startup testing. |
| 21 January – 22 January 1971 | Reactor shutdown. PM-3A Operating Report Number 60; Malfunction-71-2. |
| 22 January – 23 January 1971 | Reactor critical for startup testing. |
| 23 January 1971 | Reactor shutdown. PM-3A Operating Report Number 60; Malfunction-71-3. |
| 23 January – 24 January 1971 | Reactor critical for startup testing. |
| 24 January – 15 February 1971 | Reactor shutdown. PM-3A Operating Report Number 60; Malfunctions-71-4, 71-5, 71-6, 71-7, 71-8. |
| 15 February 1971 | Reactor critical for startup testing. |
| 15 February – 19 February 1971 | Reactor shutdown. PM-3A Operating Report Number 60; Malfunctions-71-9, 71-10. |
| 19 February – 21 February 1971 | Reactor critical. PM-3A Operating Report Number 60; Malfunctions-71-11, 71-12. |

40

| | |
|---|---|
| 21 February 1971 | PM-3A assumed McMurdo Station load. |
| 21 February – 02 March 1971 | PM-3A on the line for power operations. PM-3A Operating Report Number 60; Malfunctions-71-13, 71-14. |
| 02 March – 03 March 1971 | Reactor shutdown. PM-3A Operating Report Number 60; Malfunction 71-15, 71-16. |
| 03 March 1971 | Reactor critical. PM-3A assumed McMurdo Station load. |
| 03 March – 07 March 1971 | PM-3A on the line for power operations. |
| 18 March – 21 March 1971 | Reactor shutdown. PM-3A Operating Report Number 60; Malfunctions-71-21, 71-22, 71-23. |
| 21 March 1971 | Reactor critical. PM-3A assumed McMurdo Station load. |
| 21 March – 23 March 1971 | PM-3A on the line for power operations. |
| 23 March – 25 March 1971 | Reactor shutdown. Reactor critical. PM-3A assumed McMurdo Station load. PM-3A Operation Report Number 60; Malfunction 71-24. |
| 25 March – 24 September 1971 | PM-3A on the line for power operations producing a record run of 4400 hours 20 minutes. |
| 24 September – 01 October 1971 | PM-3A shutdown for maintenance. |
| 01 October 1971 | PM-3A assumed McMurdo Station load. |
| 01 October – 14 October 1971 | PM-3A on the line for power operations. ... |
| 14 October – 23 October 1971 | PM-3A shutdown for crew training. |
| 23 October – 28 October 1971 | PM-3A on the line for power operations. |

41

| | |
|---|---|
| 28 October – 31 October 1971 | Reactor scram. Reactor critical. PM-3A assumed McMurdo Station load. PM-3A Operating Report Number 63; Malfunction-71-48. |
| 31 October – 02 November 1971 | PM-3A on the line for power operations. |
| 02 November – 06 November 1971 | Reactor manually scrammed. PM-3A Operating Report Number 63; Malfunction-71-50. |
| 06 November 1971 | Reactor critical. Reactor manually scrammed. PM-3A Operating Report Number 63; Malfunction-71-51. |
| 07 November – 10 November 1971 | Reactor critical. |
| 10 November 1971 | PM-3A assumed McMurdo Station load. |
| 10 November – 12 December 1971 | PM-3A on the line for power operations. |
| 12 December 1971 | Reactor shutdown. |
| 13 December 1971 | Reactor brought critical. PM-3A assumed McMurdo Station load. |
| 14 December 1971 | Reactor manually scrammed. PM-3A Operating Report Number 63; Malfunctions-71-54, 71-55, 71-56, 71-57. |
| 14 December 1971 – 24 January 1972 | PM-3A shutdown for annual maintenance. |
| 24 January 1972 | Reactor critical. Reactor shutdown. Reactor critical. PM-3A Operating Report Number 64; Malfunction-72-1. |
| 25 January – 28 January 1972 | Reactor shutdown. PM-3A Operating Report Number 64; Malfunction-72-2. |
| 29 January 1972 | Reactor critical. |
| 30 January 1972 | PM-3A assumed plant load. |

42

| | |
|---|---|
| 31 January 1972 | PM-3A assumed McMurdo Station load. Reactor scrammed. Reactor critical. PM-3A Operating Report Number 64; Malfunction-72-4. |
| 01 February 1972 | PM-3A assumed McMurdo Station load. |
| 01 February – 03 March 1972 | PM-3A on the line for power operations. |
| 03 March 1972 | Reactor scram. PM-3A Operating Report Number 64; Malfunction-72-7. |
| 04 March 1972 | Reactor critical. |
| 05 March 1972 | PM-3A assumed McMurdo Station load. |
| 05 March – 19 March 1972 | PM-3A on the line for full power operations. |
| 19 March 1972 | Reactor shutdown. PM-3A Operating Report Number 64; Malfunction-72-8. |
| 19 March – 24 March 1972 | PM-3A shutdown. PM-3A Operating Report Number 64; Malfunctions-72-8, 72-9. |
| 24 March 1972 | Reactor critical. PM-3A assumed McMurdo Station load. |
| 24 March – 31 March 1972 | PM-3A on the line for power operations. |
| 31 March – 01 April 1972 | PM-3A shutdown. PM-3A Operating Report Number 64; Malfunction-72-12. |
| 02 April – 05 April 1972 | PM-3A shutdown. PM-3A Operating Report Number 65; Malfunction-72-14. |
| 05 April 1972 | Reactor critical. PM-3A assumed McMurdo Station load. |
| 05 April – 09 April 1972 | PM-3A on the line for power operations. |

| 09 April 1972 | Reactor scrammed. Reactor critical. PM-3A assumed McMurdo Station load. |
|---|---|
| 09 April -<br>19 April 1972 | PM-3A on the line for power operations. |
| 19 April 1972 | Reactor scrammed. PM-3A Operating Report Number 65; Malfunctions-72-16, 72-17. |
| 19 April -<br>28 April 1972 | PM-3A shutdown. |
| 28 April 1972 | Reactor critical. PM-3A assumed McMurdo Station load. |
| 28 April -<br>30 April 1972 | PM-3A on the line for power operations. |
| 30 April 1972 | Reactor scrammed. PM-3A Operating Report Number 65; Malfunctions-72-18, 72-19. |
| 30 April -<br>03 May 1972 | PM-3A shutdown. |
| 03 May 1972 | Reactor critical. |
| 04 May 1972 | PM-3A assumed McMurdo Station load. |
| 04 May -<br>10 May 1972 | PM-3A on the line for power operations. |
| 10 May 1972 | PM-3A shutdown. PM-3A Operating Report Number 65; Malfunctions-72-20, 72-21. |
| 10 May -<br>19 May 1972 | PM-3A shutdown. |
| 19 May 1972 | Reactor critical. PM-3A assumed McMurdo Station load. |
| 19 May -<br>20 May 1972 | PM-3A on the line for power operations. |
| 20 May 1972 | Reactor manually scrammed. Reactor critical. |

|                                    | PM-3A assumed McMurdo Station load. PM-3A Operating Report Number 65; Malfunction-72-22. |
|------------------------------------|------------------------------------------------------------------------------------------|
| 20 May – 18 September 1972         | PM-3A on the line for power operations. |
| 18 September – 16 October 1972     | PM-3A shutdown for scheduled maintenance and for Malfunction 72-26. PM-3A Operating Report Number 66; Malfunction-72-26. |
| 16 October 1972                    | Reactor critical. |
| 17 October – 22 October 1972       | PM-3A shutdown. |
| 22 October – 26 October 1972       | Reactor critical. |
| 26 October 1972 – 30 June 1973     | Reactor shutdown. PM-3A Operating Report Number 67; Appendix I. |
| 01 July 1973                       | Defueling Procedure initiated. |
| 05 July 1973                       | Defueling Procedure successfully completed. |
| 06 July – 09 October 1973          | PM-3A in Cold Iron Status. |
| 10 October 1973                    | Initiated PM-3A Removal Plan. |
| 20 October 1973                    | Final Weekly Operating Report forwarded. Henceforth Situation Reports regarding PM-3A removal will be made in accordance with the PM-3A Removal Plan of September 1973. |

## OPERATING DATA TABULATION
### 13 March 1964—30 September 1973
### CUMULATIVE OR MAX REPORTED

| Operating Data | 1964 | 1965 | 1966 | 1967 | 1968 | 1969 | 1970 | 1971 | 1972 | 1973 | TOTAL |
|---|---|---|---|---|---|---|---|---|---|---|---|
| 1. Time available for power operations (hr) | 3,145 hrs 47 min | 5,356 hrs 57 min | 6,758 hrs 19 min | 7,530 hrs 52 min | 7,470 hrs 35 min | 5,984 hrs 09 min | 5,986 hrs 14 min | 6,243 hrs 41 min | 4,706 hrs 15 min | Cold Iron | 54,132 hr 49 min |
| 2. Unscheduled Scrams | 33 | 44 | 16 | 22 | 17 | 13 | 10 | 7 | 6 | 0 | 168 |
| 3. Downtime Maintenance and Repair (hr) | 3,560 hrs 45 min | 3,427 hrs 03 min | 1,978 hrs 41 min | 1,169 hrs 39 min | 1,313 hrs 32 min | 1,871 hrs 51 min | 2,893 hrs 46 min | 2,588 hrs 19 min | 4,029 hrs 45 min | | 22,833 hr 21 min |
| 4. Peak Reactor Power (%) | NR | NR | NR | NR | NR | 116 | 117 | 110 | 98 | | 117 |
| 5. Peak Power Produced (KW) | 1500 | 1780 | 1810 | 1850 | 1900 | 1954 | 2049 | 2314 | NR | 0 | 2314 |
| 6. Peak Power Supplied (KW) | 1155 | 1435 | 1430 | 1500 | 1560 | 1540 | 1560 | 1600 | NR | 0 | 1800 |
| 7. Total Electrical Energy produced (KWH) | $3.24 \times 10^6$ | $6.94 \times 10^6$ | $8.69 \times 10^6$ | $9.52 \times 10^6$ | $1.00 \times 10^7$ | $9.7728 \times 10^6$ | $8.714 \times 10^6$ | $8.45 \times 10^6$ | $6.4816 \times 10^6$ | 0 | $7.1808 \times 10^7$ |
| 8. Total Electrical Energy exported (KWH) | $2.41 \times 10^6$ | $5.37 \times 10^6$ | $6.78 \times 10^6$ | $7.38 \times 10^6$ | $7.86 \times 10^6$ | $7.694 \times 10^6$ | $6.736 \times 10^6$ | $6.864 \times 10^6$ | $4.938 \times 10^6$ | 0 | $5.625 \times 10^7$ |
| 9. Total Core Burnup (EFPH) (Note 1) | | | | | | | | | | | |
|   Type I, Serial II | 7165.023 | NA | NA | NA | NA | NA | NA | NA | NA | NA | 7165.023 |
|   Type I, Serial I | NA | 4857.66 | 5414.24 | 4809.884 | NA | NA | NA | NA | NA | NA | 14581.78 |
|   Type II, Serial I | NA | NA | NA | 1425.656 | 6471.850 | 6430.35 | 2928.106 | NA | NA | NA | 16835.90 |
|   Type IV, Serial II | NA | NA | NA | NA | NA | 2850.725 | 2850.725 | 5454.876 | 3545.033 | NA | 10956.40 |
| 10. PM-3A Water Consumption (Gals) | 137,372 | 245,893 | 317,190 | 361,944 | 484,811 | 623,134 | 599,126 | 545,603 | 571,389 | 162,587 | 4,049,049 |
| 11. PM-3A Diesel Fuel Consumption (Gals) | | | | | | | | | | | |
|   (a) PM-3A | 40,310 | 46,444 | 18,134 | 24,614 | 51,118 | 47,624 | 37,804 | 30,779 | 45,898 | 40,050 | 382,775 |
|   (b) WD Plant | NA | NA | NA | 57,065 | 94,959 | 118,874 | 89,376 | 128,671 | 130,099 | NA | 619,044 |
| 12. Total Water Produced by Water Distillation Plant (Gals) | | | | | | | | | | | |
|   (a) Nuclear Energy | NA | NA | 1,943,432 | 2,167,768 | 2,706,529 | 1,924,261 | 2,838,496 | 1,857,171 | | | 13,494,6 |
|   (b) Diesel Fuel | NA | NA | 1,343,125 | 2,583,465 | 5,596,096 | 2,283,892 | 4,110,727 | 4,275,528 | | | 18,710,1 |

NOTE 1: Cores Type I– Serial II, Type I– Serial I and Type II– Serial I 9a, 9b, and 9c were rated at 9.51 megawatts. Core Type IV– Serial II 9d was initially rated at 9.51 megawatts but was upgraded to 11.27 megawatts.

NOTE 2: The PM-3A was in cold iron status between 18 Sep 1972 and 10 October 1973.

# SECTION III

## MALFUNCTION REPORT SUMMARY

### A. SUMMARY

There were 438 malfunctions during the period of Navy Operations, 12 March 1964 through 30 September 1973. Table III:1 has been compiled to indicate those systems which have significantly affected plant availability. Table III:1 separates malfunction reports by year and by system.

Table III:1 shows the Control Rod Drive Mechanism as a separate category in respect to the other systems. The reasoning behind this differentiation is due to the increased number of malfunctions observed in this category during the later years of plant operations. The cause of the increased malfunctions is attributal to the fact that the initial Control Rod Drive Mechanism system was a complex experimental system which was continually modified in efforts for improvement. Variable temperatures also played an important role in the breakdown of the Control Rod Drive Mechanism cables.

## TABLE III:1

### SUMMARY OF PM-3A MALFUNCTIONS FROM 12 MARCH 1964 -- 30 SEPTEMBER 1973

| Dates | Nuclear Instrumentation | Process Instrumentation | Electrical | Mechanical | CRDM | Operator Technician Error | Misc. | Total Malfunctions |
|---|---|---|---|---|---|---|---|---|
| 12 March 1964 / 02 January 1965 | 29 | 03 | 10 | 06 | 05 | 09 | 02 | 64 |
| 03 January 1965 / 01 January 1966 | 17 | 05 | 19 | 02 | 11 | 06 | 08 | 68 |
| 02 January 1966 / 31 December 1966 | 06 | 09 | 02 | 00 | 04 | 01 | 07 | 29 |
| 01 January 1967 / 30 December 1967 | 08 | 05 | 09 | 02 | 03 | 02 | 05 | 34 |
| 31 December 1967 / 04 January 1969 | 10 | 03 | 12 | 07 | 02 | 02 | 00 | 36 |
| 05 January 1969 / 03 January 1970 | 04 | 06 | 04 | 17 | 08 | 00 | 08 | 47 |
| 04 January 1970 / 02 January 1971 | 05 | 07 | 08 | 15 | 12 | 14 | 05 | 66 |
| 03 January 1971 / 01 January 1972 | 03 | 08 | 13 | 13 | 13 | 02 | 05 | 57 |
| 02 January 1972 / 30 December 1972 | 03 | 02 | 06 | 07 | 10 | 04 | 05 | 37 |
| 31 December 1972 / 30 September 1972 | | | | | | | | |
| Totals to Date | 85 | 48 | 83 | 69 | 68 | 40 | 45 | 438 |

# SECTION IV

## FINAL SUMMARY OF PM-3A HEALTH PHYSICS

### A. HPSC TYPE DEFINITIONS

1. Occurrence of an injury, as defined by OPNAVINST 5100.11 (series), to plant personnel or visitors.

2. Occurrences resulting in the exposure of personnel in excess of 350 mRem per 7 consecutive days. Occurrences resulting in the radiation exposure of personnel in excess of the quarterly limits as specified in NAVMED P-5055 or 10CFR20 shall be reported by message within 24 hours.

3. Any release of radioactivity to the environment in excess of the limits of 10CFR20.

4. Increase of radiation and/or activity levels within the plant by more than a factor of three above those normally experienced.

5. Water chemistry or radiochemistry analysis outside of a limiting condition for operation as indicated in the Operating Limits.

6. Any inability to perform a required chemistry or radiochemistry analysis not otherwise reported as a malfunction report.

7. Occurrences resulting in airborne particulate exposure to personnel greater than $3 \times 10^{-10}$ uCi/cc gross beta for any 40 hour period in 7 consecutive days.

8. Detection of airborne alpha activity greater than $2 \times 10^{-12}$ uCi/cc.

### B. THERE WERE 221 HPSC REPORTS IN THE OPERATING HISTORY OF THE PM-3A

1. Fourteen reports were Type I--outside medical assistance required. In eleven cases, the injured personnel were treated at the McMurdo Station Dispensary and returned to duty with minimal time absent from duty. In three cases, the injured individuals were admitted to the McMurdo Station Dispensary for observation. One case resulted in fourteen days lost time from duty and another case in one day lost time from duty. In the third case, the injured individual was admitted to the McMurdo Station Dispensary for observation and further evacuated to CONUS.

49

2. One Hundred twenty-three were Type II--radiation exposure in excess of 350 mRem in 7 consecutive days. Excess exposure was authorized by the OIC, PM-3A for required maintenance. Individual personnel exposure remained below the allowable quarterly limits.

3. Four reports were Type III--release of radioactivity to the environment in excess of the limits in 10CFR20. Case one occurred in 1964 and resulted from the containment purge. The isotopes present were identified as XE-133 and X-135. Although the release broke the operating limits set forth, it did not exceed the MPC. Case two occurred in 1966 and resulted from the operator erroneously dumping Hut #4 overboard. Corrective action was taken by dumping Hut #1 overboard through the same line in an effort to dilute with low activity water. Case three occurred in 1972 and resulted when 200 gallons of water at $4.6 \times 10^{-2}$uCi/ml were released to the area adjacent to the primary building south wall. Total estimated release was .035 Ci. There was no evidence of release beyond the PM-3A restricted area. The frozen material was recovered and processed for disposal. There was no increase in background radiation levels. Case four occurred in 1973 and resulted when approximately two gallons of $3.0 \times 10^{-2}$uCi/ml water leaked under the primary building loading dock due to a malfunction of the AMF radioactive waste disposal system. The surface area was disposed of and the area was smeared with results consistant with background.

4. Eleven reports were Type IV--increase in activity levels in the plant by more than a factor of three above those normally experienced. These increases in activity levels were due to various causes and as the causes were rectified the activity levels returned to normal.

6. Sixty-one reports were Type V--water chemistry or radiochemistry analysis outside of a limiting condition for operation. In each case corrective action was taken until the activity returned to the normal operating limits.

6. Five reports were Type VI--inability to perform a required chemistry or radiochemistry analysis. Each case was due to equipment being inoperative.

7. Five reports were Type VII--airborne particulate radioactivity exposure to personnel greater than $3 \times 10^{-10}$uCi/cc gross beta. Appropriate action was taken in these cases to reduce personnel exposure.

8. There were no Type VIII Reports.

## MAINTENANCE AND MODIFICATION SUMMARY

### A. WORK PROJECT SUMMARY

1. Work Project 63-4, Secondary Heating System -- Work Project 63-4 provided for improving the heating systems in the M&S Building and in the plant primary and secondary buildings. Such problems existed as a single thermostat for multiple spaces, lack of vertical air movement and unlagged piping.

2. Work Project 63-13, 02 Tank Sump Grating -- The scope of work project 63-13 covers the purchase and installation of materials in order to provide an 8 foot diameter grating cover for the 02 tank sump.

3. Work Project 63-19, Feedwater Pump Number 2 -- The scope of work project 63-19 involved the procurement of equipment and materials and the installation of an electric driven feedwater pump to replace the turbine driven feedwater pump. Specific manufacturers were specified in order to make this equipment compatible with the equipment used on feedwater pump number 1.

4. Work Project 63-25, Hands Free Monitor -- The scope of work project 63-25 provided for the procurement of materials and the installation of a "Hands Free" microphone and speaker in the 02 tank.

5. Work Project 66-1, Primary Building Addition -- Work Project 66-1 provided for the procurement and installation of a 30 by 40 foot metal frame and insulated building to house the new rad waste disposal package, provide an area for drumming of plant radioactive waste and a radioactive equipment laydown area.

6. Work Project 66-2, Vital AC-DC System Upgrade -- Work Project 66-2 provided for improving the existing vital AC-DC system. This improvement was necessitated by a requirement for additional load to be placed on the system.

7. Work Project 66-3, Reboiler -- Work Project 66-3 provided for the design, fabrication, testing and delivery of a low pressure steam reboiler unit which would consist of a complete, skid-mounted package unit, with heat exchanger, deaerating section, feed pumps, hot well, controls and instrumentation, of on-site external connections only.

8. Work Project 66-6, Topography -- Work Project 66-6 provided that a topographical survey be conducted for the area within the PM-3A control area fence.

9. Work Project 66-7, Scram Logic Drawer -- Work Project 66-7 provided for the procurement and installation of a PM-3A scram logic drawer from Bailey Meter Company. The wiring and modification of the new drawer was accomplished by the Instrument Section, Operations Branch, NNPU, Fort Belvoir, VA.

10. Work Project 66-9, Health Physics Counting Room -- Work Project 66-9 provided for a new personnel entrance into the plant so that the Health Physics Office would be isolated for use as a radiation counting room.

11. Work Project 66-10, Control of Switching Station Number 2 -- Work Project 66-10 provided for relocating switching station number 2 to the M&S Building. It was designed to improve the operation of switching station number 2 and provide the PM-3A with remote control over the heater bank feeder breakers at this switching station.

12. Work Project 66-12, Repair of CRDM -- Work Project 66-12 provided for on-site repair and modification of the control rod drive mechanisms.

13. Work Project 66-13, Pressurizer Heaters -- Work Project 66-13 provided for a modification of the existing pressurizer heaters whereby the capacity of heater bank number 1 would be 9.6 KW.

14. Work Project 66-14, Reactor Coolant Pump Bypass Switch -- Work Project 66-14 provided for modification of the reactor coolant pump control circuitry. A key lock switch was installed which would permit the reactor coolant pump to remain on, if necessary, after a reactor scram.

15. Work Project 66-15, Crud Traps -- Work Project 66-15 provided for crud traps to be installed in the pressurizer level control system and the steam generator level control system prior to entering the DP cells. The purpose of the traps was to reduce the possibility of malfunctions of/ or damage to the instrumentation.

16. Work Project 66-16, Condenser Platform Seal -- Work Project 66-16 provided for installation of materials to seal the condenser deck to prevent snow and moisture from filtering into the underfloor storage area.

17. Work Project 66-17, Manometers -- Work Project 66-17 provided for the procurement and installation of manometers to be installed across various filters in the PM-3A ventillation and air removal system.

18. Work Project 66-18, Actuator Cabinet Loss of Power -- Work Project 66-18 provided for the installation of two 480 volt relays to monitor the power to the rod actuator cabinet.

19. Work Project 66-19, Cat D-G Building Modification -- The scope of work project 66-19 included the procurement of materials and the performance of the following work on site:

a. Relocating the Allis Chalmers Diesel to the caterpillar diesel generator building.

b. Installing underground conduit and power cable from the Allis Chalmers Diesel to the switchgear in the PM-3A secondary building.

20. Work Project 66-20, Drawing Team -- Work Project 66-20 provided for updating all PM-3A as-built drawings so that these drawings represent an accurate description of the plant systems.

21. Work Project 66-21, Condensate Pump Replacement -- Work Project 66-21 provided for the procurement and installation of two condensate pumps and associated piping.

22. Work Project 66-22, Containment Purge Modification -- The purpose of Work Project 66-22 was to eliminate the problem of high containment air activity and to reduce the time required for purging the containment tanks prior to personnel entry.

23. Work Project 66-23, Pilot Operated Relief Valves -- Work Project 66-23 provided for the replacement of the pressurizer pressure relief valves with pilot operated relief valves.

24. Work Project 66-25, Rad Waste Baler -- Work Project 66-25 provided for the procurement and installation of a rad waste baler.

25. Work Project 66-26, 512 Analyzer and Typewriter -- Work Project 66-26 provided for the procurement and installation of a model ND-130 AT, 512 channel analyzer computer and an IBM computer readout typewriter.

26. Work Project 66-27, Turbine Bypass Valve -- Work Project 66-27 provided for the procurement and installation of a conversion kit to convert the subject valve to a type 476D.

27. Work Project 66-28, Main Steamline Steam Traps -- Work Project 66-28 provided for the procurement and installation of YARWAY series 130 impulse steam traps with integral strainer and blowdown valve assemblies.

28. Work Project 67-1, Secondary Building Ventilation System -- Work Project 67-1 initially was proposed for a modification to the secondary ventilation system. This work project was canceled in DEEP FREEZE 69; however materials remain on site for use as required.

29. Work Project 67-2, Primary Demineralizer Replacement -- Work Project provided for the procurement and replacement of the primary purification demineralizer with a demineralizer similar to the initial PM-3A demineralizer. This work project also provided for a modification to the present PM-3A primary coolant purification system in that the replacement demineralizer would be located outside containment.

30. Work Project 67-3, RWDS Unit Replacement - Work Project 67-3 provided for the procurement and installation of the low temperature AMF, RWDS.

31. Work Project 67-4, Source Range Shields and Dry Wells -- Work Project 67-4 initially provided for the procurement and installation of dry wells and appropriate shielding for the source range detectors. Eventually it was determined the dry wells were not feasible and there was not a necessity for the detectors to be in dry wells.

32. Work Project 67-5, Decay Heat Solenoid Valve -- Work Project 67-5 provided for the procurement and installation of a solenoid operated valve in the decay heat line which would functionally replace valve DH02-VC1.

33. Work Project 67-6, Turbine Oil Sump Vapor Extractor -- Work Project 67-6 provided for removing the existing vapor extractor, which was over capacity, from the turbine generator overhead crane structure, and installing a new lower capacity extractor on the forward manhole of the lubrication oil sump.

34. Work Project 67-7, Pressurizer Low Level Heater Cutoff Bypass -- Work Project 67-7 provided for a bypass switch to be installed on the pressurizer heater system. Initially, with two channels operating normally, a low level on either channel would cutoff the heaters.

35. Work Project 68-1, Turbine Drain Modification -- Work Project 68-1 provided for modification of the turbine drain system to consist of changing the drain tank vent connection from the hot well vent line to the exhaust trunk.

36. Work Project 68-2, Condenser Freeze Protection -- Work Project 68-2 provided for the replacement of freeze protection devices for the four main condenser air off-take and condensate return lines and the two loop seals, to include new electrical "heat tape", new insulation and a thorough check of existing breakers and wiring.

37. Work Project 68-3, Shield Water Demineralizer Relocation -- Work Project 68-3 was canceled.

38. Work Project 68-4, Low Flow Scram -- Work Project 68-4 provided for replacing the existing reactor coolant pump low power scram instrumentation system with a low differential pressure scram instrumentation system.

39. Work Project 68-5, Condensate Storage Tank -- Work Project 68-5 provided for fabrication, procurement and replacement of the condensate storage tank.

40. Work Project 68-6, Condenser Fan Bearing Replacement -- Work Project 68-6 provided for the replacement of all condenser fan bearings which had not been previously replaced.

41. Work Project 68-7, Dosing Tank Replacement -- Work Project 68-7 provided for replacing the existing dosing tank with an identical tank.

42. Work Project 68-8, Turbine Inspection -- Work Project 68-8 provided for a Class A inspection of the turbine steam path, bearings, governor system, lubricating system, and sealing system. The inspection also included the measurement of "as found" vibration and alignment and necessary correction to vibration and alignment on the reassembly and restart of the unit. The inspection was accomplished by a team of four Navy crew maintenance personnel and one Elliott Company representative who supervised the inspection.

43. Work Project 68-9, Microfilm Drawing Reproduction System -- Work Project 68-9 provided for conversion of the present blueline drawing system in the PM-3A Maintenance Office to a 35 mm microfilm system.

44. Work Project 68-10, Main Generator Neutral-Ground Resistor --
Work Project 68-10 provided for the installation of a neutral grounding
resistor between the main generator neutral and the PM-3A ground bus.
Parallel operation of the PM-3A with the McMurdo Station Diesel plant
required this installation to reduce ground fault currents at the PM-3A.

45. Work Project 68-11, Shield Water Pump Replacement -- Work
Project 68-11 deals with the procurement and replacement of the shield
water pump. The Wheeler-Economy pump was replaced with a duplicate
304 stainless steel pump.

46. Work Project 68-12, WD Plant Aux Boiler Overhaul -- Work
Project 68-12 was established for funding purposes only for procuring the
Cleaver Brooks Boiler repair parts.

47. Work Project 68-14, Reboiler Feed Pump Overhaul -- Work Project 68-14 provided for the procurement of parts for the econodyne vertical boiler feed pump overhaul.

48. Work Project 68-15, Reactor Tool Procurement -- Work Project
68-15 provided for the procurement of reactor tools for the PM-3A and
the refueling training facility at Fort Belvoir, VA.

49. Work Project 68-16, 3M Valve Recertification -- Work Project
68-16 provided for the recertification, including any rework and testing
necessary, of the PM-3A pressurizer relief valves.

50. Work Project 69-1, Shield Water System Modification -- Work
Project 69-1 recommended that the series connected reactor coolant pump
and decay heat pump cooling water jackets be reconnected in a parallel
configuration and that manually operated flow control valves be installed
downstream of each component. Shield water system cleaning eliminated
this requirement.

51. Work Project 69-2, Reboiler Proportional Feedwater Flow Control
-- Work Project 69-2 provided for the installation of a proportional level
control system to control the level in the reboiler shell. This would stabi-
lize the reboiler feedwater system and reduce the fluctuations in steam
load to the reboiler.

52. Work Project 69-3, Temporary Shield Water Storage -- Work
Project 69-3 provided for procurement of six vinyl lined swimming pools
to be used for temporary storage of shield water.

53. Work Project 69-4, Hypochlorinator Installation -- Work Project 69-4 provided for the procurement and installation of the hypochlorinator in order to provide continuous chlorination to the fresh water system.

54. Work Project 69-5, Feedwater Piping Modification -- Work Project 69-5 provided for the procurement of materials and replacement of eroded components in the feedwater system piping.

55. Work Project 69-6, Condenser Number 1 Retubing -- Work Project 69-6 provided for the retubing of condenser number 1 after the condenser was frozen.

56. Work Project 69-7, Turbine Overhaul -- Work Project 69-7 provided for the procurement of materials for a complete overhaul of the Elliott turbine-generator unit.

57. Work Project 69-8, Primary System Check Valves -- Work Project 69-8 provided for the procurement and installation of check valves in the coolant purification line and the coolant charging line.

58. Work Project 69-9, Decay Heat Pump -- Work project 69-9 provided for the procurement and replacement of the decay heat pump.

59. Work Project 70-1, Interface Modification -- Work Project 70-1 provided for modification of the PM-3A refueling tools and core handling systems. This modification was necessitated by the Core Type IV Interface Study.

60. Work Project 70-2, AMF Skid Cooling System -- Work Project 70-2 provided for increasing the cooling capacity of the AMF unit by installation of the PM-1 AMF cooler and associated pump and piping.

61. Work Project 70-3, Condenser Retubing -- Work Project 70-3 provided for the retubing of condensers 2, 3 and 4. This work project was canceled due to observations by Crew VIII personnel.

62. Work Project 70-4, Emergency Diesel -- Work Project 70-4 provided for replacing the Allis-Chalmers model 21000 emergency generator prime mover with an Allis-Chalmers Model 21000H prime mover. Included in this work project were two replacement Woodward governors for the caterpillar diesel engines.

63. Work Project 70-5, Condenser Butterfly Valves -- Work Project 70-5 was established for the purpose of replacing the condenser steam inlet valves.

64. Work Project 70-6, Condenser Stack Extensions -- Work Project 70-6 provided for the addition of extensions to the condenser stacks to eliminate wind buffeting of the stack butterfly valves and subsequent damage to the gear train.

65. Work Project 70-7, Diesel Building Heating System -- Improvements to the diesel building heating system at the PM-3A was designated Work Project 70-7. This work project was canceled and combined with Work Project 71-1.

66. Work Project 70-8, WD Plant and Diesel Building Fire Protection -- Work Project 70-8 provided for upgrading the fire protection and fire fighting capabilities in the WD Plant and Diesel Buildings.

67. Work Project 70-9, Secondary Building Pressure Relief Valves Discharge Piping -- Work Project 70-9 provided for correcting difficiencies in the PRV discharge piping in the secondary building.

68. Work Project 70-10, Turbine Oil Cooler Air Filter System -- Work Project 70-10 was established to provide an air filter system around the lower section of the oil cooler. Work project 70-1- was canceled and an alternate method of cleaning the lube oil cooler was initiated.

69. Work Project 70-11, Pressurizer Surge Line Inspection and Repair -- Work Project 70-11 provided for inspection and repair of the pressurizer surge line due to a misalignment of the piping that connects the pressurizer with the primary loop.

70. Work Project 70-12, Spent Fuel Cask Liner -- Work Project 70-12 provided for the procurement of two spent fuel cask liners.

71. Work Project 71-1, M&S Building Heating and Ventilation System -- Work Project 71-1 provided for improving the heating and ventilation system in the M&S Building. The heating problem in the emergency diesel shed was studied concurrently.

72. Work Project 71-2, Coolant Charging System Pulsation Dampener -- Work Project 71-2 provided for the installation of a pulsation dampener on the coolant charging line between the coolant charging pumps and the 02 tank.

73. Work Project 71-3, Secondary Building Noise Suppression -- Work Project 71-3 provided for the installation of noise suppression panels on

the walls and ceilings of the office spaces, control room, HP office and wardroom.

74. Work Project 71-4, Condenser Building Air Baffels -- Work Project 71-4 projected the installation of a baffel around the condensers in accordance with Hittman Associates recommendation. Work Project 71-4 was canceled in view of the lack of any significant operational problems with the PM-3A condensers during the last two years.

75. Work Project 71-5, Sonoray Caliper Ultrasonic Tester (feedwater pipe inspection) -- Work Project 71-5 provided for the procurement of a sonoray caliper thickness tester and associated spare parts for use in conducting piping inspections.

76. Work Project 71-6, Primary Loop Decon -- Work Project 71-6 provided for procuring necessary materials, establishing proper procedures and performing a decontamination of the primary loop. Work project 71-6 was modified to a limited loop decontamination.

77. Work Project 71-7, Steam Generator Tank Piping Insulation -- Work Project 71-7 provided for reinsulating the piping in the steam generator tank and repainting the inner surface of the steam generator and void tanks as necessary.

78. Work Project 71-8, Water Distillation Unit Overhaul -- Work Project 71-8 provided for overhauling water distillation units number 1 and 2. This work project was canceled due to the purchase of a third WD unit.

79. Work Project 71-9, Tritium Dilution (Brine Discharge Line) -- Work Project 71-9 provided for rerouting the brine line and salt water discharge from the WD plant in order to effectively reduce the tritium concentration in the outfall glacier.

80. Work Project 71-10, Shield Water System Modification -- Work Project 71-10 provided for removal of the spent fuel tank recirc. pump and rerouting the discharge of the shield water demineralizer into the existing suction line of the removed SFT recirc. pump.

81. Work Project 71-11, Environmental Monitoring Equipment -- Work Project 71-11 provided for the procurement and installation of environmental monitoring equipment.

82. Work Project 71-12, Turbine Bypass Switch -- Work Project 71-12 provided for the installation of a key switch on the turbine throttle valve LS-A switch box in order to allow exercising of the turbine throttle valve without tripping the turbine.

83. Work Project 71-13, Cesium Filters -- Work Project 71-13 provided for the installation of a selective cesium removal system in the radioactive waste water purification system. This cesium removal system will be in addition to the present mixed bed (H-OH) demineralizer, which although adequate for other radioactive species is not efficient for cesium atom removal.

84. Work Project 72-1, Containment Pressure Sensors -- Work Project 72-1 was based on a recommendation by the DF 71 NAVFAC Inspection Report. This report called for the installation of an additional pressure sensor in the containment closure system to provide a minimum of 1 of 2 logic for containment closure.

85. Work Project 72-2, Contaminated Water Storage -- Work Project 72-2 provided for procurement of materials and the establishing of procedures required to assemble and fill a temporary contaminated water storage system with a variable capacity of 500-600 gallons and return the contaminated water to the waste disposal system.

86. Work Project 72-3, 24 Volt Power Supplies -- Work Project 72-3 consisted of the field installation of two "eastronics" model 501-24L 24-volt DC regululated power supplies, with an external load switch, as replacements for the installed units; transferring the 115 volt AC, 60 HZ input and the regulated 24-volt DC output terminal wiring from the old unit to thenew replacement unit terminal boards; removal of the old units and unused wiring; setting the new output voltage at 24 volts and performing test RS-5 to calibrate the bistables for the new power supply.

87. Work Project 72-4, PA System Upgrade -- Work Project 72-4 provided for a general upgrading of the present PM-3A PA system. It provided that a review be conducted with a goal of increasing the capacity and audibility of the system.

88. Work Project 72-5, Tri-Carb Rehab. -- Work Project 72-5 provided for the replacement of the manual non-refrigerated tri-carb unit with a manual refrigerated tri-carb unit.

89. Work Project 72-6, Mechanical Systems Upgrade -- Work Project 72-6 was initiated to provide the PM-3A with the necessary replacement valves, flanges, fittings, pipe, supporting equipment and procedures to correct identified deficiencies and to upgrade specified portions of the secondary system.

90. Work Project 72-7, Electrical Systems Upgrade -- Work Project 72-7 provided for correcting identified deficienciencies in the electrical circuits and served as the planning document for providing and scheduling the manhours necessary to perform the essential maintenance which is beyond the capabilities of the operating crew.

91. Work Project 72-8, Instrument Systems Upgrade -- Work Project 72-8 provided for the manpower and materials to correct identified deficiencies and replace system components not replaced as part of the regular preventative maintenance program.

92. Work Project 72-9, Limited Loop Decontamination -- Work Project 72-9 provided for decontamination of the primary loop in an effort to reduce radiation levels.

93. Work Project 73-1, Emergency Core Cooling System -- The purpose of Work Project 73-1 is to design an ECCS that will mitigate the consequences of a LOCA, for installation at the PM-3A. Conceptual design and installation cost estimate for Work Project 73-1 were completed. This work project has been canceled due to the decision to decommission the PM-3A.

94. Work Project 73-2, Rad Waste Disposal PM-3A -- Work Project 73-2 provided for correcting deficiencies in radioactive waste disposal that currently exist at the PM-3A and initiating more timely shipments of waste to prevent any future backlog.

95. Work Project 73-3, Plant Systems Upgrade -- Work Project 73-3 provided for the correction of identified deficiencies in plant systems.

96. Work Project 73-6, High Temperature RWDS Evaporator Replacement -- Work Project 73-6 provided for the material and guidelines for the installation of a new high temperature RWDS evaporator along with suitable shielding, to be installed in the primary addition building.

97. Work Project 73-7, CRDM Upgrade -- The purpose of Work Project 73-7 was to determine the present material condition of the CRDM

61

system, determine if the system had been properly maintained, determine if the equipment can be repaired from available stock of spare parts and to determine what, if any, work should be done in DF 74. This work project was canceled due to the decision to decommission the PM-3A.

98. Work Project 73-8, MSS Valve Replacement -- Work Project 73-8 provided the materials and installation guidelines for replacing the existing main steam stop valve with the new Copes-Vulcan valve. The valve has been procured and is on-site. This work project was canceled due to the decision to decommission the PM-3A.

99. Work Project 73-9, Refueling Interconnect Painting -- The purpose of Work Project 73-9 was to provide suitable coating material for the welds of the refueling interconnect to prevent direct contact with the shield water and thereby inhibit corrosion activity.

100. Work Project 73-10, CRDM Cable Replacement -- The purpose of Work Project 73-10 was to provide new CRDM cables and waterproof connectors that were designed to improve the reliability of the CRDM system. This work project was canceled due to the decision to decommission the PM-3A.

101. Work Project 74-1, Turbine Inspection -- Work Project 74-1 was initiated to provide a "Class A" inspection of the turbine generator. This work project was canceled due to the decision to decommission the PM-3A.

# SECTION VI

## PM-3A PERSONNEL

### A. CHANGE OF COMMAND

The Change of Command for the PM-3A normally took place in October/ November of each year. At this time the majority of crew members were also relieved.

### B. THE CREWS IN ORDER WERE AS FOLLOWS:

   * Denotes second winter-over tour
  ** Denotes third winter-over tour

   @ Denotes second summer support tour
  @@ Denotes third summer support tour
 @@@ Denotes fourth summer support tour
@@@@ Denotes fifth summer support tour

1. Crew I Personnel, DEEPFREEZE 62

#### WINTER-OVER

| | |
|---|---|
| MITCHELL, Thomas J. | LT, CEC, USN  Officer in Charge |
| MATHERS, Winfred C. | LT, CEC, USN Plant Superintendent |
| BLACK, Albert K. | CE1 |
| BROOKS, Walter B. | SP5 |
| BRUCE, Dale R. | SKCA |
| DUBAY, Roger M. | UT1 |
| FERGUSON, Dale L. | SP5 |
| FLEMING, John P. | CECA |
| GABBERT, Raymond B. | UTCA |
| GABERLEIN, William E. | CE1 |
| KOZIKOWSKI, Stanley F. | HM2 |
| LOWE, Donald H. | SFC |
| MASCHKA, Paul R. | A1C |
| McCANN, James M. | UT1 |
| MILLER, Howard J. | UTCS |
| MILLER, Huey W. | CE1 |
| POLLOCK, Herbert W. | CECA |
| SPENCER, Sidney T. | CECS |
| STIERER, Byron A. | A1C |
| WILLIAMS, Donald O. | SP5 |

CREW I SUMMER SUPPORT

REDMAN, Bob D.                 UTCA

2. CREW II Personnel, DEEPFREEZE 63

WINTER-OVER

| | | |
|---|---|---|
| COPE, Ronald P. | LT, CEC, USN | Officer in Charge |
| MATHEWSON Maurice | LT, CEC, USN | Plant Superintendent |
| ANDREW, Robert E. | SP6 | |
| CARRIGAN, Charles E. | SK1 | |
| CARSON, Gene A. | CE1 | |
| CROWE, Edward J. | SP5 | |
| FADDEN, Dean E. | UTCA | |
| GOFF, Tommy W. | CES2 | |
| GOZA, James N. | HM1 | |
| HILSABECK, Walter G. | CMCS | |
| HOGG, Royal T. | CECA | |
| ISENHOFF, Gayle P. | SFC | |
| JOHNSON, Donnie L. | HM2 | |
| KNIGHT, Donald W. | CECA | |
| LAW, George L. | CE1 | |
| LINN, Paul E. | UT1 | |
| MILLER, Gerald J. | EOCA | |
| SMITH, Herbert G. | EO1 | |
| SOURDIFF, Lavern J. | SSGT | |
| SWINFORD, Harold D. | UTCA | |
| WAGES, Joe C. | CECA | |

SUMMER SUPPORT

| | |
|---|---|
| FEDDERSON, Bernard C. | CEC |
| GABERLEIN, William E. | CE1 |
| McCANN, James M. | UT1 |
| McKEE, Charles R. | SWF2 |
| SINGLETON, William T. | CECA |
| TOLIN, Dean | HMCA |

3. CREW III Personnel, DEEPFREEZE 64

## WINTER-OVER

| | |
|---|---|
| FEGLEY, Charles E., III | LT, CEC, USN Officer in Charge |
| BATES, Ronald G. | LTJG, CEC, USN Plant Superintendent |
| BELL, Frederick H. | HMC |
| BENDER, Neal E. | SFC |
| BERNARDO, Gerald S. | UTCM |
| CLARK, William P. | CE1 |
| COLBY, Stanley C. | UTP2 |
| CUSTEAD, Elmer B. | HMC |
| DEWEES, Bruce V. | CE1 |
| FORT, Robert E. | CECS |
| GABERLEIN, William E. | CEC * |
| GARLAND, Robert A. | CE1 |
| KAMAGAI, Takeshi | SFC |
| LODGE, David B. | HM1 |
| MOORE, Ernest H. | CEC |
| QUICK, Joe C. | SP5 |
| RANDALL, John A. | CM1 |
| REDMAN, Bob D. | UTCM |
| RISING, Harold A. | HM1 |
| SCHLOREDT, Jerry L. | CE1 |
| SCHULZ, George K. | CEC |
| SINGLETON, William T. | CEC |
| TOLIN, Dean S. | HMC |
| WOODS, Jimmy R. | SK1 |
| YOUNG, Tolbert, Jr. | SGT |

## SUMMER SUPPORT

| | |
|---|---|
| ELDRED, David T. | EON2 |
| GOLIGHTLY, Ernest J. | HM1 |
| HOLMES, William A. | CMC |
| KING, Jerry W. | CECA |
| McCANN, James M. | UT1 @ |
| McGREGOR, Leonard G. | CE1 |
| McKEE, Charles R. | SWF2 @ |
| MELTON, John L., Jr. | HMCA |
| MOOREHEAD, Jimmy R. | SW1 |
| POLLOCK, Herbert W. | CEC |
| ROMINGER, Gerald R. | CET2 |

4. CREW IV Personnel, DEEPFREEZE 65

WINTER-OVER

| | | |
|---|---|---|
| SHAFER, Willard G. | LCDR,CEC,USN | Officer in Charge |
| STEPHENSON, Walter S. | LTJG,CEC,USN | Plant Superintendent |
| BELCHER, Edgar E. | CECM | |
| BINGHAM, Ralph E. | HM2 | |
| CLARK, Richard A. | CECS | |
| DORCHUCK, Robert E. | CM1 | |
| ELDRED, David T. | EON2 | |
| EVANS, Robert L. | CEC | |
| FELTER, Paul D. | HM1 | |
| FERGUSON, Carl E. | CEC | |
| GARDNER, Keith A. | UTP2 | |
| HELMS, Henri J. | SSG | |
| HINOJOSA, Raul A. | SP5 | |
| MELTON, John L. | HMC | |
| MILLER, Frank P. | HM2 | |
| MOOREHEAD, Jimmy R. | SW1 | |
| MUCHOW, Marvin J. | UTC | |
| NELSON, Darrell L. | SK2 | |
| NOONAN, John H. | CM1 | |
| PHILLIPS, Robert C. | MSGT | |
| ROMINGER, Gerald R. | CE1 | |
| SAUNDERS, Richard S. | CE1 | |
| SPENCER, Robert F. | HMC | |
| THOMPSON, Francis S., Jr. | CE1 | |
| VIOLETTE, Ronald E. | SP5 | |
| YONKER, Chester P., Jr. | CEP2 | |

SUMMER SUPPORT

| | | |
|---|---|---|
| BERKOWITZ, Robert J. | UTB2 | |
| BROWN, Stanley | HMC | |
| FLEMING, John P. | CEC | |
| GANNON, John W. | EO1 | |
| HEEDER, Clayton A. | YN1 | |
| HOFFMANN, Edward H. | CE1 | |
| HOLMES, William A. | CMC | @ |
| HOOVER, Rex A. | CE1 | |
| McCANN, James M. | UT1 | @@ |
| McCARTHY, Jerome | CEC | |
| McGREGOR, Leonard G. | CE1 | @ |
| McKEE, Charles R. | SWF2 | @@ |
| MONOHAN, Marion L. | CEW2 | |

66

| | | |
|---|---|---|
| PLICHTA, Richard T. | HM1 | |
| POLLOCK, Herbert W. | CEC | @ |
| REED, Charles E. | CEC | |
| SCHNABEL, John G. | CE1 | |

5. CREW V Personnel, DEEPFREEZE 66

## WINTER-OVER

| | | |
|---|---|---|
| BOENNIGHAUSEN, Thomas L. | LT, CEC, USN | Officer in Charge |
| KING, Jerry W. | ENS, CEC, USN | Plant Superintendent |
| ADAMS, Melvin E. | SP5 | |
| ANDERSON, Russel F. | HM2 | |
| BELL, Myron H. | EOC | |
| BROWN, Stanley | HMC | |
| FUNKHOUSER, Earl F. | SP5 | |
| FLEMING, John P. | CECS | * |
| GANNON, John W. | EOC | |
| HAIR, Robert B., Jr. | SK2 | |
| HOFFMANN, Edward H. | CE1 | |
| HOOVER, Rex A. | CE1 | |
| MOFFAT, Robert J. | CEC | |
| O'CONNOR, George V., Jr. | CEC | |
| PERROTTI, Donald J. | SP6 | |
| PERSELL, Horace L. | HM1 | |
| PLICHTA, Richard T. | HM1 | |
| RAMSEY, Marshall E. | CMC | |
| RAY, Jerome E. | CEC | |
| REED, Charles E. | CEC | |
| ROBERTSON, James E. | CECS | |
| ROBSON, Robert J. | CEW2 | |
| WHITEMAN, Richard J. | UTCS | |

## SUMMER SUPPORT

| | | |
|---|---|---|
| ASHENDEN, Melvin C., Jr. | CEC | |
| BENEFIEL, Albert D. | SK1 | |
| BLAKE, Joseph A., Jr. | CEC | |
| GABERLEIN, William E. | CEC | @ |
| GARLAND, Robert A. | CE1 | |
| GROOVER, Edwin D. | EO1 | |
| HAMBY, Ed C., Jr. | SWCS | |
| HATFIELD, Lloyd D. | YN2 | |

| McGREGOR, Leonard G. | CE1 @@ |
| ORR, John L. | HM1 |
| PAGE, Lester D. | SFC |
| RICCIO, Theodore J. | CET2 |
| SCHLOREDT, Jerry L. | CE1 |
| STANFIELD, William D., Jr. | CEC |

6. CREW VI Personnel, DEEPFREEZE 67

WINTER-OVER

| DONOVAN, Lawrence K. | LCDR, CEC, USN | Officer in Charge |
| MILLER, Huey W. | LTJG, CEC, USN * | Plant Superintendent |
| ALLARA, V. C. | SP6 | |
| BARTLEY, John D. | CECS | |
| BERKOWITZ, R. J. | UT1 | |
| BLACK, A. K. | CEC * | |
| BLESS, J. W. | HM2 | |
| BONTEMPO, J. E. | SP5 | |
| CARSON, G. A. | CEC * | |
| COBB, R. O. | CE1 | |
| CROWSON, F. R. | SFC | |
| DOOLEY, J. E. | HM1 | |
| ERICKSON, D. L. | HM1 | |
| HEINRICHS, R. J. | EQCM | |
| HOLMES, W. A. | CMCS | |
| JAKULEWICZ, C. S. | CM1 | |
| JOHNSON, J. E. | CEC | |
| KRUPA, J. E. | SFC | |
| LINN, P. E. | WO1, CEC, USN * | |
| MONOHAN, M. L. | CE1 | |
| NELSON, D. E. | SKC | |
| ORR, J. J. | HMC | |
| POLLOCK, H. W. | CECS * | |
| RICCIO, T. J. | CET2 | |
| TWITTY, D. L. | HMCS | |

SUMMER SUPPORT

| ASHENDEN, M. C., Jr. | CEC @ |
| BENEFIEL, A. D. | SKC @ |
| BROBERG, J. G. | HMCS |
| CAVANAUGH, R. F. | CM1 |
| ESLICK, R. W. | CE1 |

68

| | |
|---|---|
| EVANS, T. R. | UTCS |
| FINE, M. L. | UTB3 |
| FORT, R. E. | WO1, CEC, USN |
| JENNINGS, M. L. | YN2 |
| McGREGOR, L. G. | CEC    @@@ |
| MILLER, G. J. | CWO2, CEC, USN |
| SCHLOREDT, J. L. | CEC    @ |
| WOOD, R. A. | CE1 |
| WYLIE, J. D. | UT1 |

7. <u>CREW VII Personnel, DEEPFREEZE 68</u>

<u>WINTER-OVER</u>

| | | |
|---|---|---|
| KOHLER, Arthur D., Jr. | LCDR, CEC, USN | Officer in Charge |
| SWINFORD, Harold D. | LTJG, CEC, USN | Plant Superintendent * |
| ALEXANDER, R. F. | CEC | |
| BROBERG, J. G. | HMCS | |
| CLARK, W. P. | CEC    * | |
| DAVISON, T. R. | CE1 | |
| DEWEES, B. V. | CEC    * | |
| ESLICK, R. W. | CE1 | |
| FORNEL, P. E., Jr. | SP6 | |
| GABERLEIN, W. E. | CEC    ** | |
| GLOSS, D. R. | SP6 | |
| GROOVER, E. D. | EOC | |
| HAUGH, J. R. | HMC | |
| MAGEE, H. J. | HM1 | |
| MARKES, J. A. | CEC | |
| McDUFFEE, J. W. | HM1 | |
| McNEISH, R. I. | SK1 | |
| METCALF, C. B. | CE1 | |
| RANDALL, J. A. | CMCS    * | |
| SCHILE, G. D. | HM1 | |
| SCHNABEL, J. G. | CEC | |
| SMITH, H. G. | EOCS    * | |
| WOOD, R. A. | CE1 | |
| YELLE, L. G. | SP6 | |

SUMMER SUPPORT

| | | |
|---|---|---|
| ASHENDEN, M. C., Jr. | CEC | @@ |
| BROOKS, W. M. | CE1 | |
| GARDNER, K. A. | UT1 | |
| HILSABECK, W. G. | CWO2, CEC, USN | |
| HOUSEL, M. D. | CMC | |
| MARTIN, D. K. | SSGT | |
| McCANN, J. M. | UTC | @@@ |
| McGREGOR, L. G. | CEC | @@@@ |
| MERCIEZ, W. R. | EO1 | |
| SINGLETERRY, D. G. | CETCN | |
| SMITH, D. L. | UT1 | |
| SWARTZ, R. D. | SK1 | |
| WRIGLEY, R. K. | PN3 | |
| ZIMMERMAN, J. L. | CM1 | |

8. CREW VIII Personnel, DEEPFREEZE 69

WINTER-OVER

| | | |
|---|---|---|
| KURTZ, James P. | LCDR, CEC, USN | Officer in Charge |
| MILLER, G. J. | CWO2, CEC, USN | Plant Superintendent * |
| LAW, G. L. | WO1, CEC, USN | * |
| ASHER, B. F. | CECS | |
| CAVANAUGH, R. F. | CM1 | |
| CHEEK, L. V. | CE1 | |
| DORCHUCK, R. E. | CMCS | * |
| ELDRED, D. T. | EOC | * |
| GARLAND, R. A. | CECS | * |
| IRBINE, A. L. | CEC | |
| MILLER, F. P. | HMC | * |
| PACE, H. C. | UT1 | |
| PUTMAN, D. W. | CEC | |
| SCHLOREDT, J. L. | CEC | * |
| SWARZ, K. | SK1 | |
| SIMMONS, J. A. | SW1 | |
| SMITH, R. M. | HMC | |
| TATE, A. C. | HMC | |
| WARD, C. A. | CEC | |
| WERNER, M. R. | UTC | |
| YOUNG, D. L. | HMC | |
| VIOLETTE, R. E. | SFC | * |

70

| BARCUS, C. K. | SP6 |
| BUCHANAN, R. J. | SP5 |
| HALES, H. L. | SFC |

SUMMER SUPPORT

| HILSABECK, W. G. | CWO2, CEC, USN @ |
| REUTTER, R. E. | CECS |
| HOUSEL, M. D. | UTC |
| FLEMING, J. P. | CEC |
| HARRIS, R. W. | PN3 |
| MELEE, T. R. | CECS |
| ROETTGER, G. C. | HM1 |
| BUNCH, D. C. | CE3 |
| DUHN, E. D. | SFC |

9. CREW IX Personnel, DEEPFREEZE 70

WINTER-OVER

| REYNOLDS, Ralph R. | LCDR, CEC, USN Officer in Charge |
| FORT, Robert E. | CWO2, CEC, USN Plant Superintendent * |
| BINGHAM, R. E. | HM1 * |
| BRANDON, J. L. | HM1 |
| BROOKS, W. M. | CEC |
| COX, R. | CECS |
| GARDNER, K. A. | UTC * |
| GUESS, T. | SK1 |
| MELEE, T. R. | CECS |
| MERCIEZ, W. R. | EOC |
| PRICE, G. L. | CE2 |
| REED, C. E. | CECS * |
| ROBERTSON, J. E. | CECS * |
| ROBSON, R. J. | CEC * |
| ROETTGER, G. C. | HM1 |
| ROGERS, B. W. | HMC |
| SINGLETERRY, D. G. | CE2 |
| SMITH, D. L. | UTC |
| WHITEMAN, R. J. | UTCS * |
| WILLIAMS, J. L. | CM1 |
| WINKLEY, D. A. | UT1 |
| ZIMMERMAN, J. L. | CM1 |
| CAISON, L. H. | SFC |
| MALCOM, D. E. | SFC |
| STRICKLIN, H. L. | SFC |

## SUMMER SUPPORT

| | |
|---|---|
| BARRETT, B. F. | SW3 |
| DUSEK, L. G. | YN2 |
| FINLAW, D. F. | CE2 |
| GROOVER, J. A. | EO1 |
| HOUSEL, M. D. | CMC    @ |
| KUKI, C. H. | SP6 |
| LOEBS, C. H. | SP6 |
| McCORMICK, T. D. | SFC |
| SELMONT, R. M. | SFC |

10. CREW X Personnel, DEEPFREEZE 71

## WINTER-OVER

| | |
|---|---|
| ARCUNI, A. A. | LCDR, CEC, USN  Officer in Charge |
| LINN, P. E. | CWO2, CEC, USN Plant Superintendent ** |
| POLLOCK, H. W. | UTCM  ** |
| ASHENDEN, M. C. | CECS |
| ANDERSON, R. F. | HM1    * |
| ANDREWS, D. L. | HMC |
| BOST, R. R., Jr. | EO2 |
| CLOPTON, R. L. | SK1 |
| CLOVER, W. B., Jr. | UT2 |
| COBB, R. O. | CEC    * |
| DOZIER, R. E., Jr. | SW2 |
| DULANEY, J. D. | HMC |
| GOODFIELD, M. C. | SFC |
| GROOVER, J. A. | EO1 |
| HARVEY, P. A. | TSGT |
| HINOJOSA, R. A. | SFC    * |
| JAKULEWICZ, C. S. | CMC    * |
| KLETT, T. F. | TSGT |
| MILLER, F. P. | HMC    * |
| NEWMAN, E. W. | CE1 |
| OBEY, R. H. | CEC |
| O'CONNOR, A. C. | CE2 |
| SCHWEIBINZ, E. R. | CE1 |
| SISK, W. A., Jr. | CE1 |
| STRAWBRIDGE, L. R. | EO2 |

SUMMER SUPPORT

| | |
|---|---|
| BURT, C. R. | CMC |
| BUSHALL, W. | CE2 |
| CARSON, G. A. | CEC |
| GROOVER, E. D. | EOC @ |
| LOEBS, C. A. | UT2 |
| NEFF, D. B. | PN3 |
| REEVES, B. G. | UT3 |
| WOOD, R. A. | CE1 @ |
| KERSHNER, M. J. | CE3 |
| BRANDT, B. K. | UTC |
| FLETCHER, D. L. | CE2 |
| SYKES, T. P. | UT1 |

11. CREW XI Personnel, DEEPFREEZE 72

WINTER-OVER

| | |
|---|---|
| BOHNING, L. R. | LT, CEC, USN  Officer in Charge |
| DORCHUCK, R. E. | EQCM  Plant Superintendent ** |
| GROOVER, E. D. | EOCS  * |
| BLAKELY, C. R. | CE2 |
| REEVES, B. G. | UT2 |
| BARRETT, B. F. | SW2 |
| McGREGOR, L. G. | CEC |
| ERICKSON, D. L. | HMC  * |
| CODY, D. J. | HM1 |
| WAHLMAN, K. M. | SP6 |
| REUTTER, R. E. | CECS |
| SYKES, T. P. | UT1 |
| PARCEL, J. E. | UT2 |
| MAINES, R. E. | MSGT |
| FINALW, D. F. | CE2 |
| MACWATTERS, R. H. | CE2 |
| JONES, E. T., Jr. | HMC |
| BUTLER, R. N. | SK1 |
| SAPHORE, V. C. | SFC |
| JONES, G. M. | HM1 |
| CONWAY, R. H. | UT1 |
| TALBERT, R. N. | SW1 |
| WESTERFIELD, W. M. | CM2 |
| WOOLDRIDGE, J. D. | CE1 |
| BRANDT, B. K. | UTC |

SUMMER SUPPORT

| | | |
|---|---|---|
| EAST, L. G. | SW1 | |
| FLETCHER, D. L. | CE2 | @ |
| GALLAGHER, W. C. | CM2 | |
| HARDING, J. N. | EO1 | |
| HORNA, D. A. | SP6 | |
| LOEBS, C. A. | UT2 | @ |
| McCARTY, D. A | SFC | |
| MONK, L. B. | SFC | |
| NEFF, D. B. | PN2 | @ |
| SCHNABEL, John G. | CECS | @ |
| SHADDIX, E. K. | CE1 | |
| TAYLOR, R. | SFC | |
| WENTZ, E. D. | SP6 | |
| YUNA, W. R. | SFC | |
| GARLAND, R. A. | CECS | @ |

12. CREW XII Personnel, DEEPFREEZE 73

WINTER-OVER

| | | |
|---|---|---|
| SCHLOREDT, J. L. | CECS | ** Officer in Charge (Acting) |
| ORR, J. J. | HMC | * |
| DELONG, D. L. | EOC | |
| CAVANAUGH, R. F. | CM1 | * |
| EAST, L. | SW1 | |
| STRAWBRIDGE, L. R. | EO1 | * |
| WELLS, D. H. | SP5 | |
| MILLER, F. P. | HMC | ** |
| MERCER, H. M. | HM1 | |
| MARQUEZ, J. J. | CE1 | |
| BUSHALL, W. | CE2 | |
| BAKER, B. | CE2 | |

PERSONNEL SCHEDULED TO WINTER-OVER WHO RETURNED
TO THE UNITED STATES AT THE END OF THE AUSTRAL SUMMER
DUE TO PLANT CONDITIONS.

| | | |
|---|---|---|
| CRANE, T. C. | LCDR, CEC, USN | Officer in Charge |
| DUBAY, R. M. | CWO3, CEC, USN | Plant Superintendent |
| ASHER, B. F. | CECS | |
| ELDRED, D. T. | EOCS | |
| FLYNN, D. F. | CMC | |

| | |
|---|---|
| HARDING, J. F. | EO1 |
| GOUGH, D. T. | SP6 |
| TURNIDGE, R. D. | SP5 |
| CARR, R. A. | SP6 |
| JOZSA, J. J. | HM2 |
| GALLAGHER, W. C. | CM2 |
| ROBSON, R. J. | CEC |
| RUMBAUGH, R. M. | SK1 |

## SUMMER SUPPORT

| | | |
|---|---|---|
| SMITH, D. L. | UTC | @ |
| ALEXANDER, R. F. | CECS | |
| HALE, R. C. | CM1 | |
| YEAZLE, C. E. | CE1 | |
| GRAFF, T. L. | SP5 | |
| DUHN, E. D. | SFC | @ |
| SNYDER, T. P. | SP6 | |
| McCARTER, I. D. | EO2 | |
| RODGERS, D. B. | UT3 | |
| OWINGS, C. M. | SP6 | |
| ZIMMERMAN, J. L. | CMC | @ |
| WORKMAN, M. W. | PN3 | |

PM-3A ENVIRONMENTAL RADIATION SURVEILLANCE PROGRAM 1972
SUMMARY REPORT

## A. Introduction

To assure the peaceful use of the Antarctic, the Antarctic Treaty was
signed in December 1959. This treaty provides that radioactive waste
shall not be disposed of in the Antarctic. Signing this agreement were the
twelve nations, including the United States, engaged in various scientific
operations in Antarctica during the International Geophysical Year. The
PM-3A Nuclear Power Plant at McMurdo was designed for complete
containment of radioactivity and for temporary storage of radioactive
waste to meet the treaty obligations and Title 10 of the U.S. Code of
Federal Regulations. To provide conclusive data that the plant would
not release activity after the scheduled startup in March 1962 background
radiation measurements were begun in December 1960.

The Antarctic Treaty, the National Science Foundation, and Title
10 of the Code of Federal Regulations set forth requirements to be
observed while operating the PM-3A. Naval Facilities Engineering
Command Instruction 11310.22A summarizes these requirements and
states that "Continual environmental surveys shall be conducted to
determine the effect on radiological background due to operation of the
plant."

## B. Background

The Environmental Radiation Surveillance Program (ERSP) at
McMurdo Station was started by the U.S. Public Health Service (USPHS)
in December 1960. The USPHS conducted the program from its inception
through October 1963. The U.S. Navy assumed the responsibility of
conducting the ERSP beginning in October 1963. Due to a reduction
in the number and frequency of samples, the full time billet for an
environmental monitor was deleted and the responsibility for conducting
the ERSP was assigned to the Health Physics and Process Control Section
supervisor, PM-3A Nuclear Power Plant, McMurdo Station. The
analysis of environmental data is performed by the Nuclear Branch,
Engineering Division, U.S. Army Engineer Power Group, Fort Belvoir,
Virginia.

C. ERSP Sampling Schedule:

The following is the current sampling schedule at McMurdo Station. (See Figure VII-1 for the location of air sampling stations).

1. Air Samples-Long-Lived beta activity

    a. Station 0101

        Location: Near Building 63 (recreation building)

        Frequency: Continuous 24-hour samples

    b. Station 0401

        Location: Cosmic Ray Laboratory (Building 84)

        Frequency: One 24-hour sample per week

2. Water and snow Samples-Long-Lived beta activity

    a. Galley water samples

        Frequency: One sample per month

    b. Seawater distillation plant distillate samples

        Frequency: One sample per week

    c. Snow samples

        Location: Mobile, reported with each sample

        Frequency: One sample per week

3. Smear test for gross contamination-gross beta activity

        Location: Various smears in McMurdo Station galley
                  and living quarters

        Frequency: Survey each week

4. Water Samples - Tritium activity

    a. Seawater samples

       Frequency: One sample per week

    b. Seawater Distillation Plant distillate samples

       Frequency: One sample per week

    c. Galley water samples

       Frequency: One sample per week

D. Analysis of Data

The data collected during calendar year 1972 was statistically compared to data collected during the years 1968, 1969, 1970, and 1971. The Duncan's Multiple Range Test[2] was used to determine whether sets of data were significantly different from each other from a statistical point of view.

A summary of the environmental radiation surveillance data collected at McMurdo Station during the years 1963-1972 is contained in tables VII-1 through VII-4. The statistical analysis of the environmental monitoring data yielded the following results.

1. Long-Lived beta activity in air samples

The data observed at all three stations during 1972 was significantly lower than that observed during 1971. Comparison with the years prior to 1971 revealed slight differences among the stations. Station 101 data was significantly lower than in 1967 while still remaining significantly higher than in 1968, 1969 and 1970. Data from stations 401 and 601 were not significantly different from 1967 and 1970 while remaining significantly higher than data observed during 1968 and 1969.

Concentrations continued at their 1971 levels through March 1972. During April they began a downward trend which resulted in levels not significantly different from the data of 1968 and 1969 being observed during October and November. During December concentrations at the stations increased significantly; however the monthly data generally was equal to or less than that observed during previous years, indicating that seasonal variation was the most probable cause. Thus, it appears that the effects of the French nuclear testing observed during 1971 have diminished.

2. Long-Lived beta activity in water.

The data for 1972 were found to be not significantly different from the data of all previous years except 1969, which was significantly higher than 1972.

3. Tritium activity in water samples.

The data from 1972 were found to be significantly lower than the 1967 data and not significantly different from the data of 1968, 1969, 1970 and 1971.

The glacier data varied over an extremely large interval with the highest measured data (taken on 7 Jan 72) being more than 3,500 times larger than the lowest data (taken on 28 July 1972). Data similar to the lowest was not observed in 1971 when the lowest data measured in the hundreds of $pCi/cm^3$.

4. Smear Samples

No significant levels of spreadable contamination were reported by the PM-3A during 1972.

E. Conclusions

The environmental radiation surveillance data as analyzed indicates no significant increase in activity in the McMurdo Station area due to the operation of the PM-3A Nuclear Power Plant.

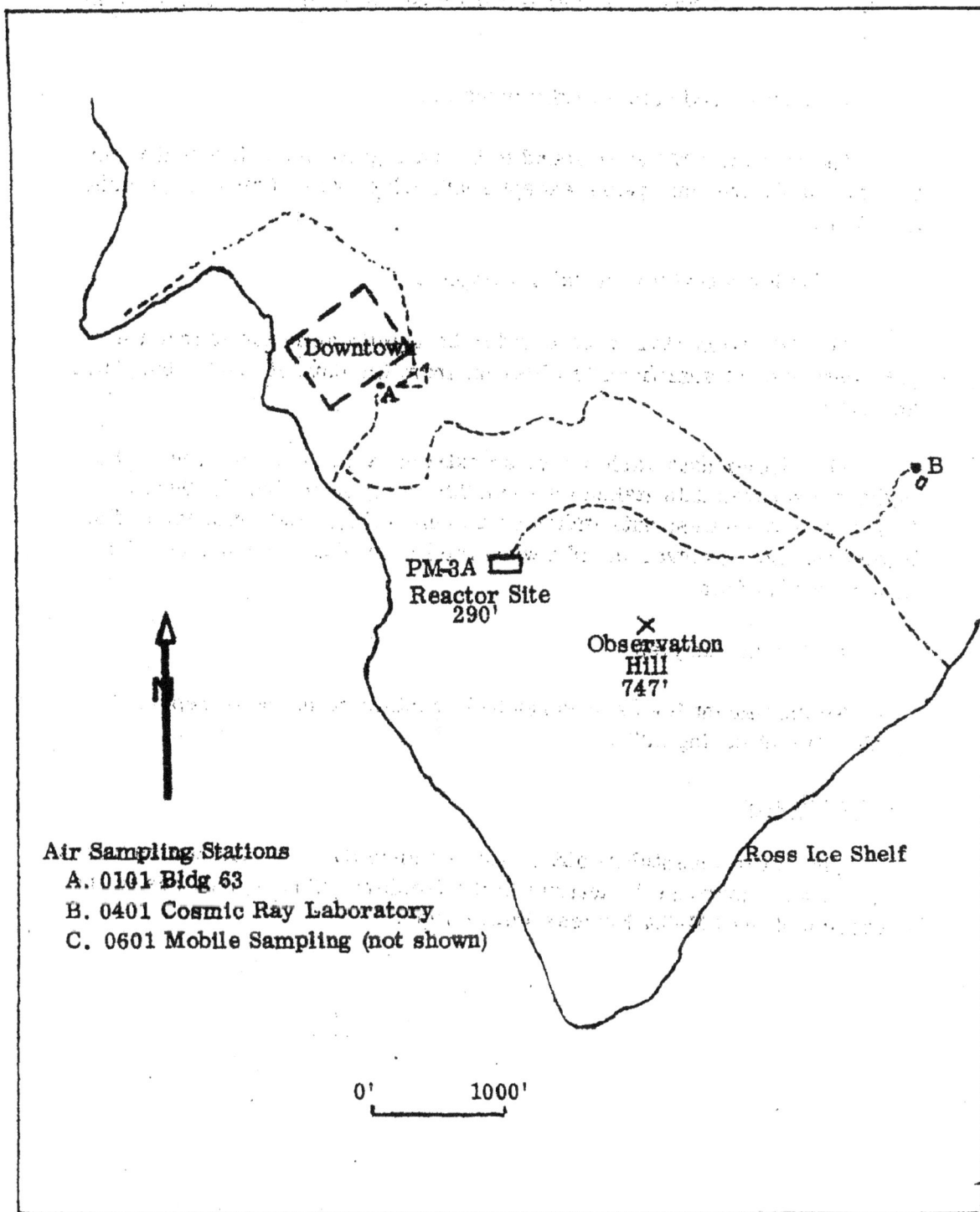

Air Sampling Stations
A. 0101 Bldg 63
B. 0401 Cosmic Ray Laboratory
C. 0601 Mobile Sampling (not shown)

ENVIRONMENTAL MONITORING SAMPLING LOCATIONS
FIGURE VII-1

## TABLE VII-1

### Mc MURDO STATION ENVIRONMENTAL DATA

Monthly Averages of Long Lived Beta Activity in air samples at McMurdo Station in Pico-Microcuries per meter cubed (pCi./M³)

| MONTH | 1963 | 1964 | 1965 | 1966 | 1967 | 1968 | 1969 | 1970 | 1971 | 1972 |
|---|---|---|---|---|---|---|---|---|---|---|
| JAN | 0.1300 | 0.0900 | 0.1119 | 0.0275 | 0.1690 | 0.0615 | 0.0503 | 0.0244 | 0.01780 | 0.1607 |
| FEB | .2000 | .0900 | .1477 | .0341 | .1574 | .0484 | .0404 | .0665 | .10923 | .1406 |
| MAR | .1600 | .0665 | .1012 | .0338 | .1496 | .0383 | .1337 | .0285 | .05383 | .1711 |
| APR | .1500 | .0711 | .0689 | .0202 | .1301 | .0353 | .1308 | .0614 | .31911 | .07603 |
| MAY | .1100 | .0800 | .0814 | .0320 | .0930 | .0411 | .0684 | .1870 | .23796 | .08753 |
| JUN | .0700 | .0800 | .0770 | .0240 | .0795 | .0352 | .0282 | .1710 | .19934 | .04644 |
| JUL | .0700 | .0857 | .0626 | .0310 | .0765 | .0306 | .0487 | .0474 | .08286 | .05692 |
| AUG | .0700 | .0590 | .0533 | .0395 | .0978 | .0341 | .0456 | .0405 | .08075 | .04306 |
| SEP | .0600 | .0770 | .0557 | .0402 | .1144 | .0399 | .0258 | .0652 | .09975 | .05910 |
| OCT | .0400 | .0989 | .0426 | .0606 | .1117 | .0382 | .0245 | .0491 | .19640 | .02854 |
| NOV | .0600 | .1194 | .0231 | .1063 | .0857 | .0458 | .0375 | .0610 | .13914 | .01913 |
| DEC | .0811 | .1193 | .0250 | .1158 | .0675 | .0338 | .0191 | .1819 | .13796 | |
| YEARLY AVERAGE | 0.1001 | 0.0859 | 0.0709 | 0.0463 | 0.1110 | 0.0406 | 0.0544 | 0.0820 | 0.13951 | 0.07867 |

## TABLE VII-2

## Mc MURDO STATION ENVIRONMENTAL DATA

Monthly Averages of Long-Lived Beta Activity in Galley Water at McMurdo Station in Picocuries Per Liter (pCi/l)

| MONTH | 1963 | 1964 | 1965 | 1966 | 1967 | 1968 | 1969 | 1970 | 1971 | 1972 |
|---|---|---|---|---|---|---|---|---|---|---|
| JAN | 25.0000 | 19.3090 | 18.3400 | 11.4915 | 0.0000 | 9.8000 | 11.6000 | 52.9000 | 6.4400 | 9.6200 |
| FEB | 4.0000 | 18.0220 | 29.4500 | 5.5700 | 0.0000 | 2.1000 | 42.8000 | 7.7000 | 4.6000 | 3.8400 |
| MAR | 12.0000 | 12.0780 | 21.8000 | 11.3860 | 8.2000 | 1.3000 | 32.4000 | 4.1300 | 4.0300 | 3.9200 |
| APR | 25.7500 | 12.7800 | 27.9100 | 10.7750 | 7.5000 | 1.3000 | 32.5000 | 4.2600 | 5.1000 | 7.8400 |
| MAY | 7.0000 | 8.6838 | 30.9700 | 7.6000 | 7.6000 | 1.3000 | 35.8000 | 8.4200 | 5.3900 | 5.8600 |
| JUN | 16.0000 | 8.1610 | 16.9300 | 5.7930 | 6.3000 | 1.3000 | 13.0000 | 2.6300 | 5.9300 | 4.9800 |
| JUL | 10.0000 | 12.1400 | 18.4900 | 11.3900 | 1.3000 | 12.0000 | 12.0000 | 4.5700 | 6.2700 | 39.0600 |
| AUG | 9.0000 | 21.5900 | 17.3000 | 6.4000 | 4.7000 | 2.0000 | 41.8000 | 7.9200 | 6.2700 | 10.3900 |
| SEP | 6.0000 | 14.8640 | 9.0200 | 7.0200 | 59.0000 | 1.2000 | 12.8000 | 2.8000 | 6.2700 | 40.0000 |
| OCT | 2.3260 | 18.7770 | 12.1800 | 1.5000 | 5.1000 | 11.0000 | 12.3000 | 17.3000 | 5.8500 | 7.4000 |
| NOV | 8.4000 | 17.8040 | 17.5750 | 0.0000 | 17.0000 | 7.3500 | 12.4000 | 4.3000 | 5.8400 | 5.7000 |
| DEC | 4.6297 | 17.2300 | 18.0930 | 0.0000 | 5.1000 | 3.7800 | 19.0000 | 4.0600 | 5.1100 | 8.5000 |
| YEARLY AVERAGE | 10.8421 | 15.1195 | 19.0048 | 6.5771 | 10.5700 | 3.6400 | 31.5300 | 10.08005 | 6.59166 | 12.2591 |

## TABLE VII-3

### Mc MURDO STATION ENVIRONMENTAL DATA

Monthly Averages of Tritium Activity in Seawater at McMurdo Station in Microcuries per liter ( U Ci/L)

| MONTH | 1967 | 1968 | 1969 | 1970 | 1971 | 1972 |
|---|---|---|---|---|---|---|
| JAN | * | 0.0080 | 0.0215 | 0.00656 | 0.00267 | 0.00187 |
| FEB | * | .0082 | .0011 | .00648 | .0022625 | .00160 |
| MAR | 0.1700 | .0080 | .0050 | .1396 | .002127 | .00351 |
| APR | * | .0082 | .0045 | ** | .004072 | .00426 |
| MAY | .0061 | .0065 | .0049 | .00801 | .00102 | .00318 |
| JUN | .0061 | .0076 | .0088 | ** | .000978 | .003045 |
| JUL | .0490 | .0055 | .0424 | .0065 | .00102 | .00520 |
| AUG | .0300 | .0045 | .0051 | .00575 | .00102 | .005015 |
| SEP | .0059 | .0042 | .0061 | .0072 | .00102 | .003432 |
| OCT | .0059 | .0044 | .0061 | .0092 | .00106 | .004225 |
| NOV | .0263 | .0510 | .0124 | .0256 | .003027 | .00330 |
| DEC | 0.0083 | * | * | 0.0027 | 0.001455 | * |
| YEARLY AVERAGE | 0.0342 | 0.0106 | 0.107 | 0.01976 | 0.00180 | 0.003513 |

*None Taken          ** Inst. Malfunction

TABLE VII-4

Mc MURDO STATION ENVIRONMENTAL DATA

Monthly Averages of Tritium Activity in Distillation
Plant Distillate in Microcuries Per Liter (UCi/L)

| MONTH | 1967 | 1968 | 1969 | 1970 | 1971 | 1972 |
|---|---|---|---|---|---|---|
| JAN | * | 0.0400 | 0.0049 | 0.00656 | 0.00228 | 0.00187 |
| FEB | * | .0091 | .0042 | .0118 | .00244 | .00257 |
| MAR | .4400 | .0480 | .0047 | .0271 | .002127 | .00351 |
| APR | * | .0160 | .0197 | ** | .004072 | .00378 |
| MAY | .0084 | .0080 | .0043 | .00801 | .00102 | .00399 |
| JUN | .0061 | .0076 | .0049 | ** | .000978 | .00264 |
| JUL | .0650 | .0055 | .0049 | .00692 | .00102 | .00413 |
| AUG | .0055 | .0045 | * | .00591 | .00102 | .00574 |
| SEP | .0031 | .0046 | .0061 | .00703 | .00102 | .00318 |
| OCT | .0200 | .0051 | .0060 | .000557 | .00106 | .00815 |
| NOV | .0130 | .0820 | .0135 | .00996 | .00409 | .00372 |
| DEC | .030 | * | 0.0174 | 0.0027 | 0.001465 | * |
| YEARLY AVERAGE | 0.0632 | 0.0209 | 0.0082 | 0.00916 | 0.00188 | 0.00348 |

*No Data Collected     **Inst. Malfunction

## SECTION VIII

## INVESTIGATION OF THE INTERCONNECT LEAK
## BETWEEN THE REACTOR (O1) TANK AND
## THE STEAM GENERATOR (O2) TANK

A. BACKGROUND

On 18 September 1972, after 2900 hours of continuous power operations, the PM-3A Nuclear Power Plant, McMurdo Station, Antarctica, was shutdown for routine maintenance. Shortly thereafter, on 19 September 1972, during a general inspection of the steam generator tank, water was discovered leaking through the normally water-tight interconnect between the steam generator tank and the reactor tank. This is the interconnect through which the primary coolant pipes pass.

Extensive testing was conducted to determine the source of the leak. Radiochemical analysis of the leak water revealed it to be from the shield water surrounding the reactor pressure vessel. By pressurizing the O2 tank, a minute crack was discovered at a weld on the insulation canning in the O1 tank on the reactor outlet leg of the primary piping. The initial leak rate was about 2.5 gallons per hour; however, subsequent partial repair of the leak reduced the leak rate to about 2.5 gallons per day. Following the repair of the crack, further pressurization tests revealed additional leak paths at both the reactor inlet and outlet shrouds.

By testing, it was established that the remaining leak paths were at points which are essentially inaccessible. The Naval Nuclear Power Unit then contracted and independent corporation to investigate the stress levels which would be produced in the vessel and piping legs if the vessel exterior was flooded. The results of this analysis indicated that a credible mechanism for such flooding existed and that the allowable stresses under ASME Code, Section 8, Division 2, would be exceeded. It was also determined that a potential chloride contamination problem existed in the Thermobestos insulation surrounding the primary coolant piping.

The contractor was then tasked with performing a detailed fatigue analysis of the vessel and appurtenances in order to determine what potential tradeoffs between fatigue and stress levels would be necessary to permit future operations and what technical approach might be taken to inspect the piping.

The initial evaluation indicated that tradeoffs could be made which would give the vessel adequate cycles for 8 to 10 years of operation, and work was started on the more complete analytical model which was expected to confirm the initial evaluation.

It was also determined that there was no easy way to perform the inspection, and that it would require many specialized organizations. Accordingly, it was decided that the contractor would supply these services by qualified subcontracts.

B.  PREPARATION FOR INSPECTION

The initial steps undertaken by the Naval Nuclear Power Unit and the contractor, after the possibility of chloride stress corrosion cracking of the austenitic stainless steel primary system piping was raised, were those necessary to confirm that all the conditions essential to chloride stress corrosion cracking were in fact, present. Since the conditions of tensile stress, high temperature, moisture and oxygen were known to be present, the only additional condition needed was the presence of chlorides. Confirmation of the presence of chlorides was obtained by an analysis of a sample of the leak water and by analysis of samples of insulation which were the same as that installed in the suspected area.

As soon as the possibility of chloride stress corrosion cracking was confirmed, arrangements for inspection of the indicated sections of piping were initiated. These sections of piping are insulated with two inch thick thermobestos insulation which is covered with 3/16 inch thick 304 stainless steel cladding and are located under approximately eighteen feet of shield water in a 300 to 400 Rem radiation field.

Due to the high radiation fields, two immediate requirements were for shielding to reduce the radiation levels and remote tooling to remove the existing canning. A full scale mock-up was constructed of the reactor pressure vessel and primary system piping in order to develop procedures for installation of the shielding and welding of the replacement insulation canning. Personnel at the PM-3A began construction of the shielding and the required rigging and support structures.

A complete set of inspection and repair procedures, to include all safety aspects and quality control assurance, was formulated. The inspection was to include chloride smear tests, visual inspection, and dye-penetrant examination of the primary piping. All defects were reported to CONUS for concurrence of the repair procedure to be utilized.

86

The inspection team arrived at the PM-3A Nuclear Power Plant, McMurdo Station, Antarctica, on 08 January 1973.

C. PM-3A INSPECTION

Upon arrival at the PM-3A, the inspection team, with the assistance of the PM-3A operating crew, began an exhaustive examination of the reactor piping. A temporary lead shield was constructed and installed in the reactor tank by the operating crew. Two sections of the insulation canning were cut open allowing a portion of the reactor pipe to be examined. It was discovered that the environment in the exposed sections contained ingredients conducive for initiating chloride stress corrosion cracking; however, no indication of the cracking was discovered on the small section of the reactor piping inspected. The pipe insulation was wet and water was found standing in one section which indicated that the insulation around the reactor pressure vessel was also wet. After analyzing the conditions found on site and conducting consultations in CONUS, the contractor recommended that the reactor not be operated until the reactor pressure vessel could be thoroughly inspected to determine if there was any evidence of chloride stress corrosion cracking.

D. NAVAL NUCLEAR POWER UNIT RECOMMENDATION

Acting on the Contractor's recommendation, and in view of the estimated cost ($1,290,000), time (26 months), and exposures (40-60 Rem) involved; the lack of an operational requirement; and the uncertainties involved in accomplishing the inspection and repair; the Officer in Charge, Naval Nuclear Power Unit recommended to the Commander, Naval Facilities Engineering Command that the PM-3A be removed.

# SECTION IX

## PM-3A REMOVAL OPERATION

### A. GENERAL

Article V of the Antarctic Treaty states:

1. Any nuclear explosions in Antarctica and the disposal there of radioactive waste material shall be prohibited.

2. In the event of the conclusion of international agreements concerning the use of nuclear energy, including nuclear explosions and the disposal of radioactive waste material, to which all of the Contracting Parties whose representatives are entitled to participate in the meetings provided for under Article IX are parties, the rules established under such agreements shall apply in Antarctica.

Thus the PM-3A must be dismantled and shipped to the Continental United States for disposal.

### B. IMPLEMENTATION

Due to the severe Antarctic climate during the winter months, the removal effort will be principally carried out during the austral summer, i.e., from October through February. It is anticipated that at least three summer seasons will be required to remove the PM-3A.

The project will be divided into two phases annually: planning and execution. The planning phase will extend from March through September of each year until the project is completed. This work will be performed at the Naval Nuclear Power Unit, Fort Belvoir, Virginia. During this time, Activity Specifications, Detailed Working Procedures and drawings will be developed. Materials, tools and equipment required during the forthcoming austral summer will be procured and shipped to Antarctica. Training and plant familiarization courses will also be conducted during this time.

The execution phase will extend from October through February of each year until the project is completed. The work will be performed by personnel of the Naval Nuclear Power Unit at the PM-3A, McMurdo Station Antarctica, under the command of an Officer in Charge. The PM-3A

Officer in Charge reports to the Officer in Charge, Naval Nuclear Power Unit, Fort Belvoir, Virginia.

During the austral winter after the first removal season (March-October 1974), the PM-3A will be manned by a nine man crew. The crew will consist of a Chief Petty Officer in Charge, Health Physicist and seven operating and maintenance personnel. The primary duties of this crew will be to operate and maintain the water distillation plant and continue the environmental monitoring program. Additionally, the crew will operate and maintain essential plant systems such as heat, electrical power and radiation monitoring. These systems will be required for the second execution season.

In addition to operating and maintaining essential systems, the crew will also provide plant security. No fissile material will be on site; however, large quantities of radioactive waste will remain. Since the site will be manned continuously, access control will be maintained to prevent inadvertent exposure to ionizing radiation by unauthorized personnel.

Non-contaminated plant equipment which is compatible with existing McMurdo Station utility systems will be transferred to Commander, Naval Support Force, Antarctica, for further utilization. The remaining non-contaminated equipment will be shipped to a Defense Property Disposal Agency in the Continental United States for salvage or disposal, as required. Contaminated equipment will be shipped to CONUS for disposal as radioactive waste.

80.887—Ft Belvoir

www.ingramcontent.com/pod-product-compliance
Lightning Source LLC
Chambersburg PA
CBHW050620110426

42813CB00010B/2621